Levees and Dams

Juan Lorenzo · William Doll
Editors

Levees and Dams

Advances in Geophysical Monitoring and Characterization

Springer

Editors
Juan Lorenzo
Department of Geology and Geophysics
Louisiana State University
Baton Rouge, LA, USA

William Doll
East Tennessee Geophysical Services, LLC
Oak Ridge, TN, USA

Introduction by Öz Yilmaz

ISBN 978-3-030-27369-9   ISBN 978-3-030-27367-5   (eBook)
https://doi.org/10.1007/978-3-030-27367-5

This Springer imprint is published by the registered company Springer Nature Switzerland AG
The registered company address is: Gewerbestrasse 11, 6330 Cham, Switzerland

Introduction

Only a third of the world's great rivers remain free-flowing—just 90 of the 246 rivers more than 1000-km-long flow without interruption. The world's rivers are interrupted by dams and levees, which constitute critical components of the infrastructures of most nations in the world. They serve indispensable functions—irrigation, water supply, flood control, electric generation, and recreation. Safe operation and maintenance of dams and levees are crucial for both sustaining these functions, and avoiding potential disaster and loss of life. Moreover, a substantial number of dams and levees in many countries are nearing the end of their life spans—requiring close monitoring of their structural safety.

Storm surge barriers of the Netherlands and New Orleans are two of the most extreme engineering works in the world. Much of the landmass of the Netherlands has been reclaimed from the North Sea by levees and dams built over the past two thousand years. The Delta Works in the Netherlands is the largest flood protection project in the world. This project consists of 13 surge barriers. The Oosterscheldekering is the largest surge barrier in the world—9 km long. The dam is based on 65 concrete pillars with 62 steel doors, each 42 m wide. It is designed to protect the Netherlands from flooding from the North Sea. The Maeslantkering is a storm barrier with two movable arms—when the arms are open, the waterway remains an important shipping route and when the arms close, a protective storm barrier is formed for the city of Rotterdam. Closing the arms of the barrier is completely automated without human intervention.

The Great Wall of Louisiana is a storm surge barrier constructed near the confluence of and across the Gulf Intracoastal Waterway and the Mississippi River Gulf Outlet near New Orleans. The barrier runs generally north–south from a point east of Michoud Canal to the Bayou Bienvenue flood-control structure. Navigation gates on the barrier reduce the risk of storm surge coming from Lake Borgne and the Gulf of Mexico.

Every four years, the American Society of Civil Engineers (ASCE) issues a report card for the American infrastructure. The report card depicts the condition and performance of American infrastructure in the familiar form of a school report card—assigning letter grades based on the physical condition and needed investments for improvement. The 2017 ASCE grade for levees and dams is D—a cause for concern and a call for action. The nationwide network of levees in the USA is more than 30,000 miles. As development continues to extend into flood-plains along rivers and coastal areas, an estimated $80 billion is needed in the next 10 years to maintain and improve the nation's system of levees. There exist more than 90,000 dams in the country with an average age of 56 years. With an increase in population and thus development, the overall number of high-hazard potential dams has increased—with the number climbing to nearly 15,500 in 2016. It is estimated that it will require an investment of nearly $45 billion to repair aging, high-hazard potential dams.

Geophysical methods are indispensable to characterize the near-surface formation prior to planning and design of dams and levees, and monitoring their structural integrity during their lifetime. This volume is devoted to case studies for investigation of seepage risk and monitoring structural safety of dams and levees. In recent years, various types of fiber-optic sensors have enabled accurate and efficient structural monitoring in civil and geotechnical engineering. The fiber-optic technology is especially suitable for monitoring large or elongated structures, such as dams, dikes, levees, bridges, and pipelines.

The first chapter in this volume, entitled "Statistical Estimation of Soil Parameters in from Cross-Plots of S-Wave Velocity and Resistivity Obtained by Integrated Geophysical Method" by Hayashi et al., describes the application of an integrated geophysical and geotechnical borehole data analysis to derive cross-plots of S-wave velocity and resistivity and various geotechnical parameters for Japanese levees. Cumulative length of the geophysical survey line traverses is nearly 670 km on 40 rivers in Japan. The geotechnical borehole data were collected from about 400 boreholes located along the geophysical survey line traverses.

The second chapter in this volume, entitled "Application of Seismic Refraction and Electrical Resistivity Cross-Plot Analysis: A Case Study at Francis Levee Site" by Wodajo et al., describes a case study to assess the integrity of earthen embankment at the site affected by sand boil formations during the 2011 Mississippi River flood event. Results from seismic refraction and electrical resistivity surveys conducted at the Francis Levee site indicate seven distinct anomalies that might be associated with seepage. Specifically, using the seismic velocity and electrical resistivity values of the anomalies on the waterside as limiting values, a cross-plot analysis was performed to identify similar anomalies on the landside. The results indicate that preferential flow occurs within the sand layer in an old oxbow.

The third chapter in this volume, entitled "A Borehole Seismic Reflection Survey in Support of Seepage Surveillance at the Abutment of a Large Embankment Dam" by Butler et al., describes installation of a modern monitoring instrumentation at the Mactaquac Generating Station, a 660-MW hydroelectric facility located on the Saint John River—approximately 20 km upriver from Fredericton, New Brunswick, Canada. The objective of this study was to confirm the location of the steeply inclined interface between an embankment dam and a concrete diversion sluiceway as accurately as possible for installing seepage monitoring instrumentation. Specifically, installation of a fiber-optic distributed temperature sensing (DTS) cable as close as possible to the sub-vertical contact between the concrete diversion sluiceway and the clay till the core of the adjacent zoned embankment dam required an accurate knowledge of the dam's internal structure. Because of lack of detailed as-built drawings, a seismic reflection survey was conducted along a sub-parallel borehole, offset by approximately 1 m at the surface and by an estimated 4 m at the dam's foundation at a depth of 50 m. A wall-locking seismic tool with eight receivers was used in two different orientations to capture P- and S-wave reflections from the concrete–clay interface. Based on the S-wave image, which helped delineate the concrete–clay interface, two 50-m-long boreholes for seepage monitoring instrumentation was installed within an estimated 50 cm of the interface.

The fourth chapter in this volume, entitled "Self-potential Imaging of Seepage in an Embankment Dam" by Bouchedda et al., describes a case study to investigate seepage in Les Cèdres embankment dam in Valleyfield, Canada, by integrating self-potential tomography (SPT), electrical resistance tomography (ERT), thermometry, electromagnetic (EM) conductivity, and magnetic measurements. SPT consists of inverting self-potential data to retrieve the source-current density distribution associated with water flow pathways in embankment dams. The embankment dam is used to channel water from the Saint Lawrence River to a hydroelectric plant. The SPT inversion utilizes the resistivity model of the dam, which is obtained by ERT. EM conductivity maps allowed identifying two linear anomalies caused by metal-shielded electrical cables. The magnetic survey shows an important anomaly zone that is probably related to a metallic object. The SPT shows a few seepage locations on the upstream dam side at a depth interval of 4–5 m. Two of these seepages were confirmed by geotechnical testing. All observable seepage outlets on the downstream side can be related to the SPT anomalies and are observed as conductive zones in the resistivity model.

The fifth chapter in this volume, entitled "Optical Fiber Sensors for Dam and Levee Monitoring and Damage Detection" by Inaudi, describes the use of optical fiber sensors for monitoring dams and levees to detect damaged locations. Case studies for the surveys with various types of optical fiber sensors include (1) a water reservoir in Spain with plastic membrane to detect leaks through the membrane and the perimeter levee; (2) Nam Gum rockfill dam in Laos with concrete face where to detect leaks through the concrete plinth; (3) Luzzone concrete arch dam in Switzerland to monitor temperature evolution during concrete setting; (4) some levees in Louisiana to monitor movements between wall panels to detect anomalies and impending panel failure; (5) an earthen levee in the Netherlands to detect early signs of levee failure; (6) a river dam in Latvia with a hydropower plant to detect leaks across bitumen joints; (7) sinkholes affecting rail and road structures in Kansas to detect impending sinkhole formation; (8) embankment dam with clay core in Spain to monitor deformation of the clay core; (9) Val de la Mare reservoir in Jersey Island with mass concrete dam wall to monitor deformations induced by alkali silica reaction in concrete; and (10) El Mauro mining tailing dam in Chile to monitor long-term deformations and pore pressure.

The sixth chapter in this volume, entitled "Application of the Helicopter Frequency Domain Electromagnetic Method for Levee Characterization" by Smiarowski et al., presents two case studies using a HEM system for levee characterization and hazard detection at Retamal Levy, Rio Grande Valley in Texas and the flood-control levees of Sacramento Valley in California. Airborne remote sensing systems, such as HEM, can be deployed to survey large areas required by levee characterization. The HEM involves towing an electromagnetic transmitter and receiver that measure signals proportional to the electrical conductivity of the ground. The HEM provides electrical conductivity information about the earth from about the top 1 to 100 m below surface. Data are typically transformed to apparent conductivity, which removes variations in system altitude and allows easier interpretation of ground material. For levee characterization, the HEM-derived conductivity mapped in 3D gives an indication of the geometry of sand channels and clay layers. In one of the case studies presented, the HEM data enabled detection of sandy channels and delineation of their spatial extent, including old oxbows and buried river channels that provide seepage pathways under the levee, which may cause sand boils or levee collapse from foundation erosion. In the second case study, high-resistivity values from the HEM data indicated dry, sandy conditions, and led to the discovery of significant cracking in the levee due to desiccation of the levee material.

Given the fact that levees and dams serve indispensable functions, including irrigation, water supply, flood control, electric generation, and recreation, safe operation and maintenance of dams and levees are crucial for both sustaining these functions and avoiding potential disaster and loss of life. The papers included in this volume demonstrate the successful application of geophysical methods to monitor the structural safety of levees and dams.

Urla   Öz Yilmaz
May 2019

Contents

Statistical Estimation of Soil Parameters in from Cross-Plots of S-Wave Velocity and Resistivity Obtained by Integrated Geophysical Method

Koichi Hayashi, Tomio Inazaki, Kaoru Kitao and Takaho Kita

Abstract Cross-plots of S-wave velocity and resistivity obtained by geophysical methods statistically estimated geotechnical soil parameters, Fc, $D20$, blow counts, and the soil types, of levee body and foundation for Japanese levees. The S-wave velocity and the resistivity were collected from surface wave methods and resistivity methods respectively. Total survey line length of the geophysical methods was about 670 km on 40 rivers in Japan. The Fc, $D20$, blow counts, and soil types were collected from about 400 boring logs carried out on geophysical survey lines. S-wave velocity and resistivity at the depth of the blow counts were extracted from two-dimensional geophysical sections. The total number of extracted data, blow counts and soil type, was about 4000. The data was grouped by levee body and foundation. A polynomial approximation estimated the soil parameters from S-wave velocity and resistivity. A least squares method optimized the coefficients of the equation. Accuracy of the estimation was statistically evaluated by comparing estimated and actual soil parameters. The correlation coefficients between estimated and actual parameters ranged between 0.43 and 0.8. The polynomial approximations with the optimized coefficients calculated soil parameter sections from S-wave velocity and resistivity sections.

K. Hayashi (✉)
OYO Corporation, San Jose, USA
e-mail: khayashi@geometrics.com

Geometrics, Inc., San Jose, USA

T. Inazaki
PWRI Tsukuba Central Institute, Tsukuba, Japan

K. Kitao
CubeWorks, Tsukuba, Japan

T. Kita
TK Ocean-Land Investigations, Nishinomiya, Japan

© Springer Nature Switzerland AG 2019
J. Lorenzo and W. Doll (eds.), *Levees and Dams*,
https://doi.org/10.1007/978-3-030-27367-5_1

1

Introduction

Conventional levee assessments use invasive borings which provide useful and detailed information of levees. However, borings are expensive and cannot provide continuous information along a levee in heterogeneous environments. Non-invasive, rapid and spatially continuous investigation methods are needed to supplement traditional investigation techniques. Many researchers have been trying to apply geophysical methods to levee investigations (e.g. Dunbar et al. 2007). Surface wave methods (e.g. Ivanov et al. 2006) and resistivity methods (Liechty 2010) are often applied to such investigations because S-wave velocity and resistivity obtained by these methods are very valuable to estimate the soil condition of levees.

Both S-wave velocity and resistivity, however, reflect many physical properties and do not directly relate to engineering soil parameters such as cohesion, internal friction angle, grain size distribution, and permeability. We proposed an integrated geophysical method (Hayashi et al. 2009; Inazaki et al. 2009) to evaluate levee soil condition quantitatively. The proposed method mainly consists of the surface wave method using a land streamer and the resistivity method using capacitively-coupled resistivity equipment. The cross-plots of the S-wave velocity and the resistivity estimate the soil condition of levees in the method.

Geotechnical soil parameters, such as soil type (clay, sand or gravel), fine fraction content (Fc) and grain size ($D20$: diameter at which 20% of the sample's mass is comprised of particles with a diameter less than this value), are particularly important information for levee safety evaluation from an engineering point of view. Many engineering analysis methods such as slope stability, seepage flow, subsidence and liquefaction analyses use these soil parameters. In these types of analyses, the soil parameters are obtained by borings and laboratory tests. Geophysical properties obtained through the geophysical methods, such as S-wave velocity and resistivity, do not directly relate to the soil parameters. For that reason, geophysical methods have not been widely used for levee safety assessment. Several researchers have been trying to theoretically estimate the soil parameters from the geophysical properties in terms of a rock physics theory that is increasing in popularity in oil and gas exploration (Konishi 2014). In this paper, we estimate the geotechnical soil parameters, Fc, $D20$, blow counts, and the soil type, in terms of a statistical approach using geophysical and geotechnical data collected from a Japanese levee. The collected data in this study will play an important role in the theoretical study as well.

This paper summarizes the integrated geophysical method, introduces a database storing the results of geophysical investigation, borings logs, and laboratory characterization of samples from the boring logs, describes a statistical estimation of soil parameters using cross-plot analysis of S-wave velocity and resistivity, and shows an application example at a Japanese levee.

S-Wave Velocity and Resistivity in Levee Investigation

Seepage and erosion, shear strength and soil types are examples of primarily important factors that must be used to evaluate the safety of levees. We will review the relationship between geophysical properties (S-wave velocity and resistivity) and geotechnical soil parameters (shear strength and soil types) in this section.

S-wave velocity is directly related to shear modulus which is particularly important to levee assessment. Small strain shear modulus (G_0) is a function of the S-wave velocity (V_S), according to:

$$G_0 = V_S^2 D \qquad (1)$$

where, D is material density. It is well known that the S-wave velocity is mainly affected by shear stiffness or porosity. A considerable number of studies have been made on the correlation between the S-wave velocity and shear strength (e.g. Imai and Tonouchi 1982). Figure 1 shows a correlation between S-wave velocity and N-value (blow counts) obtained from standard penetration tests (SPT : JIS 2005) at many Japanese levees with soil classification. S-wave velocities in Fig. 1 were obtained by a surface wave method performed on the levee surface. The black line in Fig. 1 is a regression line obtained by the least squares method. It is clear that the N-value increases as the S-wave velocity increases although there is large scatter.

Resistivity is a function of many physical properties such as porosity, pore fluid resistivity, water saturation, and grain size distribution. The conductivity (inverse of resistivity) of a porous medium is expressed by an equation as follows (Imamura et al. 2007):

$$\sigma_R = \frac{1}{a} \cdot \phi^m \cdot S^n \cdot \sigma_W + \sigma_C \qquad (2)$$

where, σ_R is conductivity of medium, σ_W is pore fluid conductivity, σ_C is conductivity due to clay minerals, ϕ is porosity, S is water saturation, a, m and n are constants. The first term on the right-hand side of Eq. (2) is known as Archie's equation and if fluid resistivity and water saturation are constant, resistivity is a function of porosity and resistivity decreases as porosity increases. A second term on the right-hand of Eq. (2) is the effect of clay minerals included in soils. It is well known that the effect of the second term cannot be neglected and may be dominant in saturated clayey unconsolidated soils. We may, therefore, reasonably conclude that the resistivity of soils mainly correlates to soil type. Figure 2 shows the example of correlation between resistivity and effective grain size ($D20$). Although there is large scatter, we can see that grain size increases and soil type is changing from clay to sand and gravel as resistivity increases.

Figure 3 (left) shows a schematic relationship between S-wave velocity and resistivity. The S-wave velocity indicates shear stiffness or degree of compaction and resistivity indicates soil type as mentioned above. Figure 3 (right) shows a schematic relationship between geophysical properties, S-wave velocity and resistivity, and the

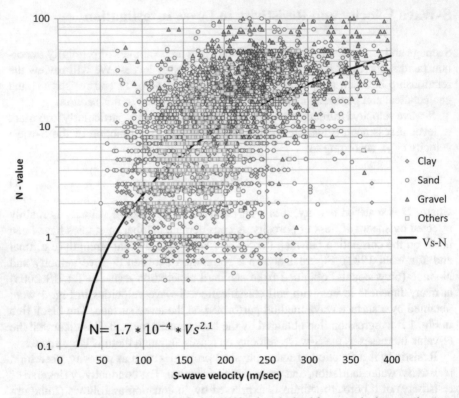

Fig. 1 Correlation between S-wave velocity and N-value obtained from standard penetration tests at many Japanese levees with soil classification. S-wave velocities were obtained by a surface wave method performed on the levee surface. The black line is a regression line obtained by the least squares method

vulnerability of levees. Loose and sandy levees are more dangerous compared with tight and clayey levees. Permeability is one of the most important parameters for levee safety assessment. It mainly relates to grain size distribution, such as clay or sand, and degree of compaction (Creager et al. 1944) although many other factors have an effect on the permeability. As mentioned above, the degree of compaction relates to shear modulus, and grain size distribution relates to resistivity. Through this it may be possible to qualitatively estimate the permeability from S-wave velocity and resistivity. Figure 3 shows the concept of cross-plots of S-wave velocity and resistivity on levee safety assessment. This implies that the geophysical methods can be used to evaluate the safety of levees.

However, Fig. 3 is quite qualitative and a more quantitative interpretation is required to apply the geophysical methods to levee safety assessment from an engineering point of view. Both S-wave velocity and resistivity reflect many physical properties. They do not directly relate to engineering parameters such as shear stiffness and permeability as well as other soil parameters such as grain size distribution. We have applied an analysis method in which soil parameters (*Fc*, *D20*, blow counts

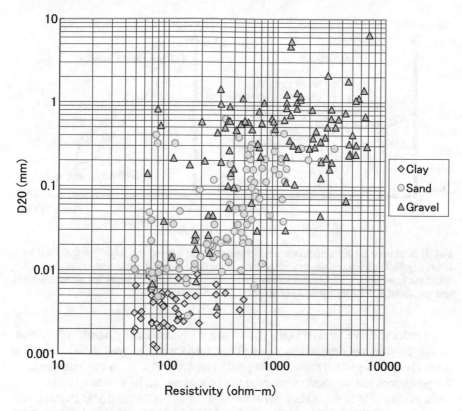

Fig. 2 Correlation between resistivity and effective grain size (*D20*) obtained from laboratory tests at many Japanese levees with soil classification

and soil type) are statistically estimated. Then we used a cross-plot of S-wave velocity and resistivity in order to apply the integrated geophysical methods to levee safety assessment more quantitatively.

Surface Wave Method

Surface waves (Rayleigh wave) are elastic waves propagating along the ground surface and their energy concentrates beneath the ground surface. The velocity of surface wave propagation strongly depends on S-wave velocity of the ground. If a subsurface S-wave velocity varies with the depth, a propagating velocity varies with its frequency or its wavelength. This characteristic is called dispersion. A surface wave method is a seismic method in which sub-surface S-wave velocity structure is estimated by the analysis of the dispersive character of the surface waves (e.g. Nazarian et al. 1983; Park et al. 1999).

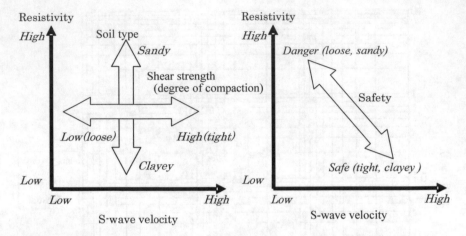

Fig. 3 Schematic relationship between geophysical properties and soil condition (left) and levee safety (right). The left figure shows a schematic relationship between S-wave velocity and resistivity. The right figure shows a schematic relationship between geophysical properties, S-wave velocity and resistivity, and the vulnerability of levees

In order to move receivers quickly, we use a land streamer (Inazaki 1999) comprising 24–48 geophones on aluminum plates, respectively, aligned in series at 1–2 m intervals by two parallel ropes on the ground surface (Fig. 4). In the land streamer, the geophones are not stuck in the ground surface and can be moved quickly.

In the analysis of the surface wave method, a CMP (Common Mid Point) cross-correlation (CMPCC) analysis (Hayashi and Suzuki 2004) is applied to waveform data firstly and a multi-channel analysis of surface-waves (MASW) developed by Park et al. (1999) is applied secondly. The CMPCC analysis is applied to raw shot gathers and CMPCC gathers are calculated in order to improve lateral resolution of S-wave velocity profiles. The MASW is applied to each CMPCC gather so that

Fig. 4 A geophone on an aluminum plate of land streamer used in a surface wave method. The land streamer comprises 24–48 geophones on aluminum plates, respectively, aligned in series at 1–2 m intervals by two parallel ropes on the ground surface

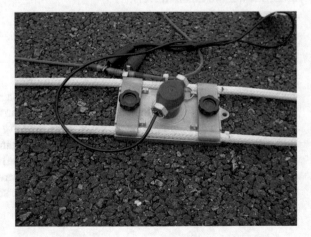

dispersion curves are constructed. A non-linear least squares method (Xia et al. 1999) is applied to each dispersion curve for reconstructing a one-dimensional (1D) S-wave velocity model. We obtain a pseudo two-dimensional (2D) S-wave velocity section by aligning the 1D S-wave velocity models.

Capacitively-Coupled Resistivity Method

A capacitively-coupled resistivity (CCR) method (Groom 2008) is an alternative resistivity method in which capacitors are used as electrodes. Unlike a conventional resistivity method, the CCR method does not use metallic stakes and enables us to measure the resistivity of the ground very quickly. The fundamental principle of capacitive coupling is that AC current will pass through a capacitor. In a CCR instrument, a cable (or metal plate) acts as one half of a capacitor, while the earth functions as the other half. This cable-earth capacitor has a variable capacitance depending on the earth conditions, but an AC current generated by the transmitter will pass from the cable into the ground. At the receiver, the transmitter-generated ground current will generate an AC voltage that is coupled into the CCR receiver and measured. The CCR receiver is conceptually equivalent to an AC Volt meter.

The Geometrics' OhmMapper (Fig. 5) is used as a CCR instrument in our investigations. The OhmMapper uses shielded twisted-pair cables as line sources and receivers in contrast with a traditional galvanic resistivity method which uses metallic stakes as point sources and receivers. A dipole-dipole array is used in the OhmMapper. The transmitter drives a 16.5 kHz signal into the cable shield and that signal is coupled to the ground through the capacitance of the cable. We have applied the method to many site investigations (Yamashita et al. 2004) and have come to the conclusion that the method enables us to delineate a precise resistivity image very quickly at least down to a depth of 5–10 m depending on site conditions.

Fig. 5 Data acquisition of capacitively-coupled resistivity (CCR) method using Geometrics' OhmMapper

Database of S-Wave Velocity and Resistivity

In order to develop statistical analysis, the results of the integrated geophysical method performed for levee soil investigations were collected and stored in the database (Hayashi et al. 2014).

We collected the results of surface wave methods and resistivity methods performed at 40 Japanese rivers as well as the results of borings performed on survey lines of geophysical methods. Total survey line length of the surface wave and resistivity methods was about 670 km and the number of borings was about 400. We generally used a land streamer (Inazaki 1999) and a capacitively-coupled resistivity (CCR) method (Groom 2008) in the data acquisition of the surface wave methods and the resistivity methods respectively. The surface wave method and the resistivity method were performed at the crest or toe of the levees. Analyzed results were saved as standard XML format defined by SEGJ (Hayashi et al. 2012) and stored in a web-based database for subsequent analysis.

Relationships between S-wave velocity, resistivity, Fc, $D20$, blow counts (N-value) obtained from the standard penetrating tests (SPT) and soil types were collected and stored in the database. The S-wave velocity and the resistivity at the depth of the blow counts were extracted from 2D geophysical sections. The depth of borings is generally less than 20 m. The total number of extracted data was about 4000. The soil type was classified as clay, sand and gravel for the sake of simplicity. Unusual soil types, such as organic clay, loam or weathered rocks etc. were rejected before the analysis. The data was grouped by levee body and foundation. The number of data points in the levee body and foundation were 560 and 3485, respectively. The data in the levee body and foundation can be considered as unsaturated and saturated soil above and below the ground water level respectively.

The results of laboratory tests, Fc, $D20$ etc. associated with borings were also collected and stored in the database. They were stored together with corresponding S-wave velocity and resistivity that were mainly measured by laboratory tests or loggings and not extracted from geophysical methods performed on the ground surface. The number of laboratory test data was about 1000. The results of laboratory tests were not grouped by levee body and foundation.

Statistical Estimation of Geotechnical Soil Parameters

Cross-Plot of S-Wave Velocity and Resistivity

Figure 6 shows correlation between S-wave velocity and resistivity at 40 Japanese levees with soil classification. The S-wave velocity and resistivity were obtained from a surface wave method and a capacitively-coupled resistivity (CCR) method performed on crest or toe of levees. Data shown in Fig. 6 were sampled at the depth where soil type could be confirmed by soil samples obtained with SPT.

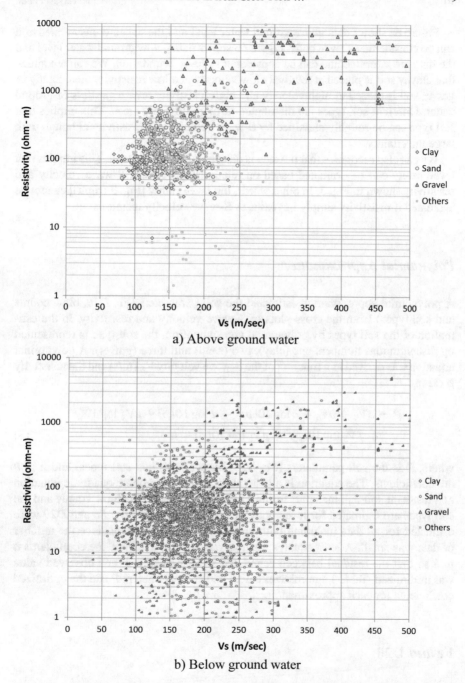

a) Above ground water

b) Below ground water

Fig. 6 Correlation between S-wave velocity and resistivity at 40 Japanese levees with soil classification. **a** Data above ground water level and the data are unsaturated and can be considered as levee body. **b** Data below ground water level and the data are saturated and can be considered as levee foundation

Figure 6a shows data above ground water level and the data are unsaturated and can be considered as levee body. Figure 6b shows data below ground water level and the data are saturated and can be considered as levee foundation. We can recognize that clayey soil is placed at relatively low velocity and low resistivity area. Sandy or gravel soil is placed at high velocity and high resistivity area regardless of ground water. Different soil types are distributed through different areas. This implies that soil type can be roughly classified by S-wave velocity and resistivity although with large uncertainty.

We can also recognize that the data is distributed in a wide area. This implies that levee safety can be evaluated based on relationship between S-wave velocity and resistivity shown in Fig. 3. In other words, the levee safety [Fig. 3 (right)] cannot be evaluated if resistivity simply increases as S-wave velocity increases.

Polynomial Approximation

A polynomial approximation estimated the soil parameters (Fc, D20, blow counts and soil types) from the cross-plots of S-wave velocity and resistivity. In the estimation of the soil types by polynomial approximation, the soil type is represented by discontinuous numbers one (clay), two (sand) and three (gravel). A polynomial equation was derived as a function of the S-wave velocity V_S (m/s) and the resistivity ρ (Ω m):

$$P = aV_S^2 + bV_S + c \log 10(\rho)^2 + d \log 10(\rho) + eV_S^2 \log 10(\rho)$$
$$+ fV_S \log 10(\rho)^2 + gV_S \log 10(\rho) + h \tag{3}$$

where, P is the soil parameters (Fc, D20, blow counts and soil types) and a to h are coefficients. The equation is a general form of a bi-variable quadratic equation and resultant soil parameters P forms a quadratic surface. The blow counts and the soil types are estimated for levee body or foundation separately. Fc and D20 were estimated for all data including both levee body and foundation since the number of data was small. A least squares method was used to calculate the coefficients a to h so that the residual between calculated soil parameters P and observed value was minimized. Table 1 summarizes calculated soil parameters P and the optimized coefficients for each approximation.

Fc and D20

Figure 7 shows the distribution of the soil parameters, Fc and D20 as a function of S-wave velocity and resistivity calculated from Eq. (1) with coefficients shown in Table 1. In Fig. 7, the symbols are Fc or D20 obtained by laboratory tests and

Table 1 Calculated soil parameters P, optimized coefficients for each approximation, and correlation coefficients between observed and calculated parameters

Soil parameters		Fc	D20	Blow counts (N)		Soil type	
				Levee body	Foundation	Levee body	Foundation
Obtained parameter (P)		0–100	$\text{Log}_{10}(D20)$	$\text{Log}_{10}(N)$	$\text{Log}_{10}(N)$	1 (clay) to 3 (sand)	1 (clay) to 3 (sand)
Unit		%	mm	Times	Times		
Coefficients	a	0.0008008	−0.0000063	−0.0000319	−0.0000024	−0.0000062	−0.0000002
	b	−0.4031928	0.0010764	0.0081824	0.0063255	−0.0072263	0.0019388
	c	−13.15444	0.3117605	0.2270993	−0.1177596	0.5333744	0.0938575
	d	37.55066	−1.394286	−1.038734	0.0137909	−1.527523	−0.5366671
	e	−0.000133	−0.0000029	0.0000082	−0.0000044	0.0000016	−0.0000064
	f	0.0403905	−0.000805	−0.0005265	0.0006654	−0.0025515	0.000198
	g	−0.0854496	0.0054981	0.0013658	0.0004906	0.0111545	0.0032458
	h	67.97225	−0.8842084	0.8181751	−0.141405	1.711534	1.406812
Correlation coefficients		0.43	0.44	0.74	0.55	0.80	0.54

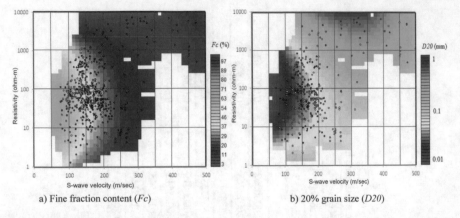

a) Fine fraction content (*Fc*)   b) 20% grain size (*D20*)

Fig. 7 Distribution of the soil parameters, *Fc* (**a**) and *D20* (**b**) as the function of S-wave velocity and resistivity calculated from the polynomial equation with coefficients shown in Table 1. The symbols are *Fc* or *D20* obtained by laboratory tests and the color of the symbols represents the observed value of *Fc* or *D20*. The colored areas in the plot represent *Fc* or *D20* calculated by the polynomial approximations

the color of the symbols represents the observed value of *Fc* or *D20*. The colored areas in the plot represent *Fc* or *D20* calculated by the polynomial approximations. We can recognize that *Fc* decreases and *D20* increases as the S-wave velocity and the resistivity increase from bottom-left to top-right although there is large scatter in data obtained from laboratory tests. It indicates that soil type changes from clay to gravel with the S-wave velocity and the resistivity increase. It should be noted that the number of data used in the prediction of *Fc* and *D20* were smaller than one of blow counts and soil types. More data is necessary to increase the accuracy and reliability of the prediction.

Blow Counts

Figure 8 shows the distributions of the blow counts in the same manner as the *Fc* and *D20* mentioned above. The reason for large scatter in the data is mainly due to a difference of area or volume of measurements as discussed later. Unlike the *Fc* and *D20*, S-wave velocity and resistivity data used in the approximations were obtained from geophysical methods performed on the ground surface. In the levee body (left), the blow counts are a function of both S-wave velocity and resistivity. They increase with an increase in the S-wave velocity and the resistivity. In contrast, color boundaries are almost vertical in the levee foundation (right). This implies that the blow counts are mainly a function of the S-wave velocity beneath the ground water level. This is because the resistivity beneath the groundwater level mainly reflects resistivity of ground water and is not sensitive to soil type.

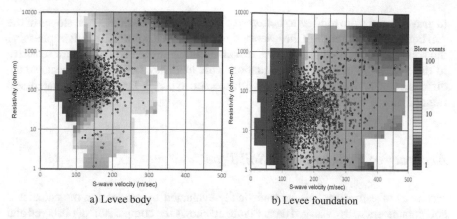

a) Levee body   b) Levee foundation

Fig. 8 Distribution of blow counts (*N*-value) in levee body (**a**) and foundation (**b**) as the function of S-wave velocity and resistivity. The symbols are blow counts obtained by standard penetration tests and the color of the symbols represents the observed value of blow counts. The colored areas in the plot represent blow counts calculated by the polynomial approximations

Soil Type

Figure 9 shows the distribution of soil types in the same manner as the blow counts mentioned above. In Fig. 9, color shading represents the value of the soil type and blue, yellow and orange colors correspond to one (clay), two (sand) and three (gravel) respectively. We can recognize that the soil types change (increases) from clay (1.0) to gravel (3.0) with an increase in S-wave velocity and resistivity from bottom-left to top-right. This agrees with the distribution of collected data shown as color symbols. Approximate boundaries of soil types (shown as broken lines), clay to sand and sand

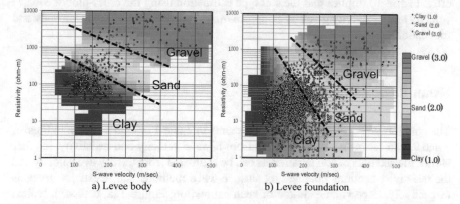

a) Levee body   b) Levee foundation

Fig. 9 Distribution of soil type in levee body (**a**) and foundation (**b**) as the function of S-wave velocity and resistivity. Color shading represents the value of the soil type and blue, yellow and orange colors correspond to one (clay), two (sand) and three (gravel) respectively

to gravel, trend from top-left to bottom-right. It should be noted that the slopes of the soil boundaries are gentle in the levee body and steep in the foundation. It implies that the soil type is more sensitive to the resistivity in the levee body and more sensitive to the S-wave velocity in the foundation. This tendency is reasonable because the difference of resistivity associated with the soil types is large in unsaturated soil rather than saturated soil.

Accuracy of Estimation for Soil Type

Accuracy of estimation can be statistically evaluated by comparing the calculated soil parameters with observed data. Figure 10 shows the comparison of observed and calculated soil parameters, Fc, $D20$ and blow counts. The correlation coefficients between observed and calculated parameters are summarized in Table 1. We can recognize that calculated soil parameters are generally consistent with observed ones although there was a large scatter. The correlation coefficients imply that the polynomial approximation reasonably predicts the soil parameters from S-wave velocity and resistivity.

Figure 11 shows the comparison of soil types. Data were grouped into four groups (1.0–1.5, 1.5–2.0, 2.0–2.5, 2.5–3.0) by the calculated soil type. In each group, the numbers of actual soil types (clay, sand, gravel) were counted and shown as proportioned in Fig. 11. The proportion can be considered as probability of estimation. For example, if the calculated soil type is smaller than 1.5 in the levee body, two-thirds of data are classified to be clay and one-third are classified to be sand. If the calculated soil type is larger than 2.5, 90% of data are classified to be gravel regardless of being from the levee body or foundation. It is clear that as the value of calculated soil type increases, the probability of sand and gravel increases although there was a large error. Figure 10 implies that the soil type estimation using the cross-plot of S-wave velocity and resistivity gives us an approximate soil structure of the levee body and foundation.

Example of Estimation

The polynomial equation (1) with coefficients in Table 1 or charts shown in Figs. 7, 8 and 9, can estimate soil parameters from S-wave velocity and resistivity. In other words, the sections of soil parameters can be estimated from the S-wave velocity and the resistivity sections obtained by surface wave methods and resistivity methods respectively. Here is an example of such estimation. Figure 12a, b show a typical example of the S-wave velocity and the resistivity sections obtained at a levee in Ibaraki prefecture, Japan. Figure 13a shows cross-plots of S-wave velocity and resistivity obtained from sections shown in Fig. 12 in the levee body (left) and foundation (right) respectively. Color indicates soil type calculated by the polynomial equation as shown in Fig. 9. We can see that the distribution of levee body and foundation

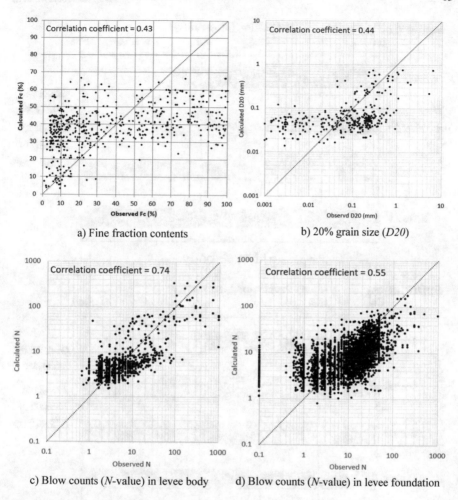

a) Fine fraction contents

b) 20% grain size (*D20*)

c) Blow counts (*N*-value) in levee body

d) Blow counts (*N*-value) in levee foundation

Fig. 10 Comparison of observed and calculated soil types, Fc (**a**), D20 (**b**) and blow counts (**c** and **d**)

data are different. Figure 13b shows a soil type section obtained from the cross-plot shown in Fig. 13a. We can recognize that a levee body is generally more clayey than a foundation at a distance range between 200 and 1100 m. It is consistent with the soil type confirmed at Boring A that shows sand at the levee body and gravel at the levee foundation. In the foundation, the end of the section (1100–1300 m) is more clayey than the rest of the section. The estimated soil type is generally consistent with the soil type observed by borings.

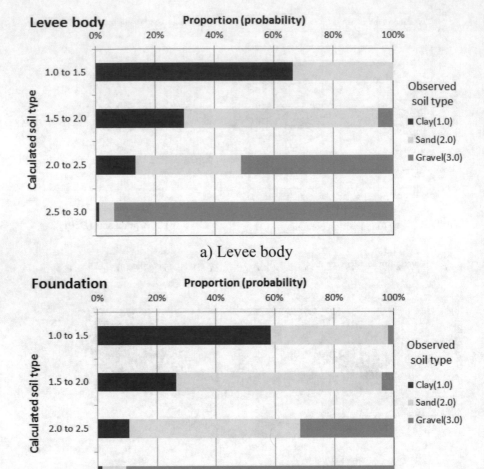

a) Levee body

b) Foundation

Fig. 11 Comparison of estimated and actual soil type. Data were grouped into four groups (1.0–1.5, 1.5–2.0, 2.0–2.5, 2.5–3.0) by the calculated soil type. In each group, the numbers of actual soil types (clay, sand, gravel) were counted and shown as proportion that can be considered as probability of estimation

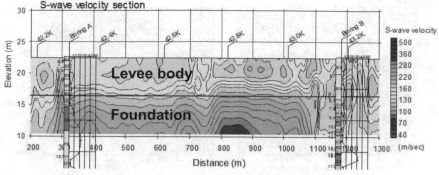

a) S-wave velocity section obtained from surface-wave method

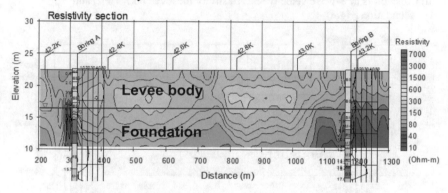

b) Resistivity section obtained from capacitively-coupled resistivity (CCR) method

Fig. 12 Typical examples of S-wave velocity and resistivity sections obtained at Ibaraki prefecture, Japan. Boring logs (Boring A and Boring B) performed on the survey line are also shown. Blow counts are shown at the righthand side of boring logs. Colors on boring logs indicate soil types and light blue, pale yellow, and orange represents clay, sand, and gravel respectively. High S-wave velocity and low resistivity at a distance of 300 m correspond to a manmade structure. Horizontal change of S-wave velocity and resistivity at a distance of 1100 m corresponds to a geological boundary. The levee is typical levee in Japan and was performing well when survey was done

Discussion

As mentioned before, there is generally large scatter in correlations between geophysical properties and soil parameters. The scatter makes engineers hesitate to apply geophysical methods to quantitative levee safety assessment.

The main reason for the large scatter is the difference of area or volume of measurements. Geotechnical in situ tests, such as borings or soundings, and laboratory tests generally measure soil parameters of very small volumes (several centimeters) of ground. In contrast, the geophysical methods average a large area or volume (tens of centimeters to several meters) of ground. The laboratory tests are more accurate and precise than geophysical methods if we compare a very small portion of

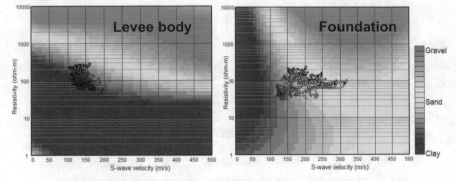

a) Cross-plots of S-wave velocity and resistivity for levee body (left) and
   foundation (right)

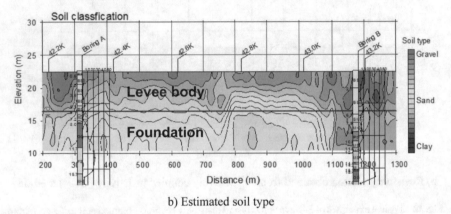

b) Estimated soil type

Fig. 13 Example of soil type estimation. **a** Cross-plots of S-wave velocity and resistivity obtained
from sections shown in Fig. 12 in levee body (left) and foundation (right) respectively. Color
indicates soil type calculated by the polynomial equation as shown in Fig. 9. **b** A soil type section
obtained from the cross-plot shown in (**a**)

the ground. The laboratory tests, however, only represent a very small portion of
the ground rather than the whole structure and only provide spatially discontinuous
information. In contrast, geophysical methods can average a large area or volume of
the ground and provide continuous information.

It should be noted that the geophysical properties and soil parameters have fun-
damentally different physical meaning. The geophysical properties are generally a
simple physical property as Eq. (1) for example. In contrast, soil parameters are
generally expressed as functions of many physical properties. There is no direct relationship between geophysical properties and soil parameters and it causes the scatter
in the correlations.

Another reason for the scatter is the non-uniqueness of geophysical processing.
Surface geophysical measurements generally do not provide geophysical properties
of the ground directly and data processing, called inversion, is generally used to

estimate the geophysical model from observed data. The inversion tries to find a geophysical model that can explain observed data in the data processing. A problem in inversion is that several different models can explain observed data. The problem is called non-uniqueness and causes uncertainty in the results and scatter in the correlations.

Resistivity of pore fluid in Eq. (2) can also play an important role. Variation of pore fluid resistivity may contribute significantly to the scatter of the estimation. Measuring the pore fluid resistivity may increase the accuracy of soil parameter estimation.

The paper described the use of multiple geophysical methods and joint interpretation of multiple geophysical properties. The correlation coefficients imply that the methods will be able to predict soil parameters quantitatively despite the existence of large scatter. The authors are thinking that the combination will play an important role in levee safety assessment. Figure 3 illustrated that a single geophysical property does not directly relate to levee safety. Quantitative assessment of levee safety requires multiple geophysical properties and the method presented in the paper demonstrated the effectiveness of joint interpretation of multiple geophysical properties.

In conclusion, laboratory tests, boring or sounding, and geophysical method have their own advantages and disadvantages. It is important to combine or integrate different methods together rather than compare accuracy and resolution of single methods. It should be noted that the study shown in the paper only uses data from the 670 km of Japanese levees and the result cannot be simply applied other levees. Collecting investigation results from other levees and accumulating the relationship between geophysical properties and soil parameters are very important.

Conclusions

The geotechnical soil parameters, Fc, $D20$, blow counts, and soil types of levee bodies and foundations were statistically predicted using the cross-plots of S-wave velocity and resistivity. The S-wave velocity was obtained from a surface wave method and the resistivity was obtained from a capacitively-coupled resistivity method. Data were collected from 670 km of geophysical methods and 400 borings. A polynomial approximation was used to estimate the soil parameters from geophysical properties. Accuracy of estimation was statistically evaluated and the correlation coefficients between estimated and actual parameters ranged between 0.43 and 0.8. The results imply that the geophysical properties obtained by geophysical methods, such as the S-wave velocity and the resistivity, can be used not only for qualitative interpretation but also quantitative engineering analyses.

References

Creager, W. P., Justin J. D., and Hinds, J., 1944, Engineering for Dams, Volume 1, New York, Wiley.

Dunbar, J. B., Smullen, S., and Stefanov, J. E., 2007, The use of geophysics in levee assessment, Symposium on the Application of Geophysics to Engineering and Environmental Problems, 2007: 61–68.

Groom, D., 2008, Common misconceptions about capacitively-coupled resistivity (CCR) what it is and how it works, Proceedings of the symposium on the application of geophysics to engineering and environmental problems 2008, 1345–1350.

Hayashi, K. and Suzuki, H, 2004. CMP cross-correlation analysis of multi-channel surface-wave data. Exploration Geophysics, 35, 7–13.

Hayashi, K., Inazaki, T., and SEGJ Levee Consortium, 2009, Integrated geophysical investigation for safety assessment of levee systems (Part 1): methodology, process and criterion for the safety assessment, Proceedings of the 9th SEGJ International Symposium–Imaging and Interpretation, 2009: 118.

Hayashi, K., Inazaki, T., Takahashi, T., and Digital Standard Format Consortium of SEGJ, 2012, Proposal for a standard digital file format of geophysical sections for civil engineering investigations in Japan, Proceedings of the symposium on the application of geophysics to engineering and environmental problems, 2012: 9.

Hayashi, K., Takahashi, T., Inazaki, T., Kitao, K., and Kita, T., 2014, Web-based database of integrated geophysical method for levee safety assessment, Proceedings of the symposium on the application of geophysics to engineering and environmental problems, 2014.

Imai, T. and Tonouchi, K., 1982, Correlation of N-value with S-wave velocity and shear modulus, Proceedings of the second European symposium on penetration testing, 67–72.

Imamura, S., Tokumaru, T., Mitsuhata, Y., Hayashi, K., Inazaki, T., and SEGJ Levee Consortium, 2007, Application of integrated geophysical techniques to vulnerability assessment of levee, Part4, comparative study on resistivity methods, Proceeding of the 116th SEGJ Conference, 120–124 (in Japanese).

Inazaki, T., 1999, Land Streamer: A new system for high-resolution S-wave shallow reflection surveys, Proceedings of the symposium on the application of geophysics to engineering and environmental problems, 1999: 207–216.

Inazaki, T., Hayashi, K., and SEGJ Levee Consortium, 2009, Integrated geophysical investigation for safety assessment of levee systems (Part 2): acquisition and utilization of ground truth data, Proceedings of the 9th SEGJ International Symposium–Imaging and Interpretation, 2009: 134.

Ivanov, J., Miller, R. D., Stimac, N., Ballard, R. F., Dunbar, J. B., and Steve Smullen, S., 2006, Time-lapse seismic study of levees in southern New Mexico, SEG Technical Program Expanded Abstracts, 2006: 3255–3259.

JIS A1219, 2005, Method for standard penetration test, Japanese Industrial Standards.

Konishi, C., 2014, Cross-plot analysis by using rock physics-based thresholds for an evaluation of unsaturated soil, SAGEEP, 149.

Liechty, D., 2010, Geophysical surveys, levee certification geophysical investigations, DC resistivity, Symposium on the Application of Geophysics to Engineering and Environmental Problems, 2010: 103–109.

Nazarian, S., Stokoe, K. H., and Hudson, W. R., 1983, Use of spectral analysis of surface waves method for determination of moduli and thickness of pavement system: Transportation Research Record 930, 38–45.

Park, C. B., Miller, R. D., and Xia, J., 1999, Multimodal analysis of high frequency surface waves: Proceedings of the symposium on the application of geophysics to engineering and environmental problems '99, 115–121.

Xia, J., Miller, R. D., and Park, C. B., 1999, Configuration of near surface shear wave velocity by inverting surface wave: Proceedings of the symposium on the application of geophysics to engineering and environmental problems 1999: 95–104.

Yamashita, Y., Groom, D., Inazaki, T., and Hayashi, K., 2004, Rapid near surface resistivity survey using the capacitively-coupled resistivity system: OhmMapper, Proceeding of the 7th SEGJ International Symposium, 292–295.

Application of Seismic Refraction and Electrical Resistivity Cross-Plot Analysis: A Case Study at Francis Levee Site

Leti T. Wodajo, Craig J. Hickey and Thomas C. Brackett

Abstract Geophysical methods such as electrical resistivity tomography (ERT) and seismic refraction tomography (SRT) provide a rapid, more economical, and non-invasive option of investigation of dams and levees with better and more complete sub-surface coverage. Several factors, such as water content and porosity, affect both SRT and ERT results although their sensitivity may differ. By combining electrical resistivity and seismic refraction tomography, a unique assessment of the integrity of earthen embankment dams and levees can be obtained. Cross-plot analysis based on seismic and electrical attributes to seepage and piping can be used to achieve this goal. In this study, a method of combining SRT and ERT using cross-plot analysis is discussed. The method is applied to geophysical surveys conducted at the Francis Levee Site, a site affected by sand boil formations during the 2011 Mississippi river flood event. Requiring consistency between seismic velocity and electrical resistivity models to predict feasible porosity values, an anomaly on the waterside that could be associated with the sand boil formations is identified. Using the seismic velocity and electrical resistivity values of the anomaly on the waterside as limiting values, a cross-plot analysis is performed to identify similar anomalies on the landside. The results from the geophysical methods, cross-plot analysis, and with the help of the geomorphology of the site, indicate that preferential flow occurs within the sand layer in an old oxbow. Sand boils at the site outcrop where the overlying clay layer is thin or the weakest.

Introduction

The 2011 flood report by the Mississippi Levee Board identified as many as twelve areas associated with seepage within the state of Mississippi (Nimrod 2011). The Francis Levee site is one of the locations affected by the flood (Fig. 1).

L. T. Wodajo (✉) · C. J. Hickey
National Center for Physical Acoustics, University of Mississippi, University, MS, USA
e-mail: ltwodajo@olemiss.edu

T. C. Brackett
Viceroy Petroleum, Houston, TX, USA

© Springer Nature Switzerland AG 2019
J. Lorenzo and W. Doll (eds.), *Levees and Dams*,
https://doi.org/10.1007/978-3-030-27367-5_2

During the 2011 flood event, three main sand boils were observed and mitigated by the construction of sandbag berms (Nimrod 2011). After the first sand boil, green dot on Fig. 2, was mitigated, two more sand boils, red dots on Fig. 2, surfaced approximately 90 m landward in the field. After the initial mitigation, the US Army Corps of Engineers (USACE) extended the berm of the levee and constructed 16 relief wells.

The Francis Levee site is located in the Mississippi River flood plain, which is composed of Holocene and Pleistocene-aged meander deposits formed by the migration of the Mississippi River across its floodplain. The fluvial depositional environments include point bars, channel-fill deposits, natural levees and back swamp deposits (Saucier 1994).

Saucier (1994) notes that the convex portion of meander bends typically hosts point bar deposits near the channel with overbank/back swamp deposits occurring further away. As the distance from the meander increases, it is expected that deposits will decrease in grain size (Brackett 2012). The three sand boil formations at the Francis Levee Site fall within an old channel or an oxbow (Fig. 3) .

Cross section A-A′ in Fig. 2 indicates that the levee is underlain by a clay-rich overburden averaging 3 m thick (Fig. 4). Below the clay overburden, the sediment coarsens into silt, and eventually a thick sand unit, which is assigned to the permeable substratum (Brackett 2012). Similar cross section is observed on the waterside of the levee with sand layer connecting the waterside to the landside. With adequate water pressure, water can flow from the waterside to the landside through the permeable sand substratum.

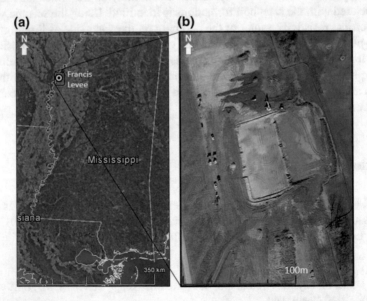

Fig. 1 **a** Francis levee site (34° 5′ 9.48″N, 90° 51′ 52.56″W) located 0.8 km west of Francis (Google Earth 2015), **b** aerial photography taken during mitigation of the levee (Google Earth 2013)

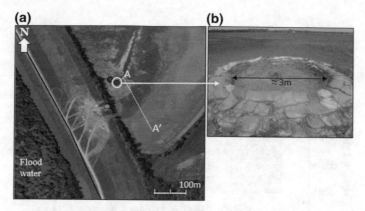

Fig. 2 **a** Location of the three sand boils (Nimrod 2011), and **b** mitigation of sand boil with sand bags (Nimrod 2011)

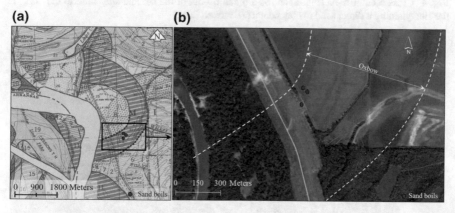

Fig. 3 **a** ancient courses of the Mississippi River reconstructed from multiple aerial photographs (Fisk 1944). The different colors indicate the stage of the river at the time. Meander belt around the Francis Levee site is indicated by the broken blue lines, **b** relative location of the sand boils (red dots) and meander belt edges (broken white line) (Google Earth 2015)

A possible model for the sand boil formations is preferential flow through the coarser grained sands that filled an old oxbow (Fig. 5). This coarse-grained high permeability sand layer acts as a flow channel for sub-surface seepage connecting the waterside to the landside. High water level on the waterside due to flood events generates enough hydraulic head to initiate seepage. This seepage then appears on the surface as a sand boil where the upper confining clay layer is the weakest.

Geophysical methods, such as seismic refraction and electrical resistivity provide a rapid, more economical, and non-invasive option of investigation with better and more complete sub-surface coverage. These methods and others have been extensively used separately and/or in combination for identify existing internal problems with dams and levees (Dahlin and Johansson 1995; Hickey et al. 2009, 2015; Kim

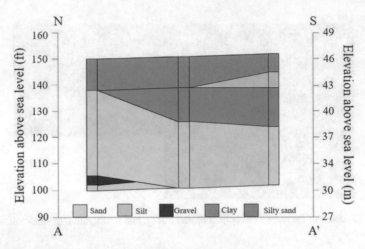

Fig. 4 Cross section A-A' (Fig. 2) shows a pinching-out of the silt and silty sand to the north of the site, yielding a direct sand to clay contact (Brackett 2012)

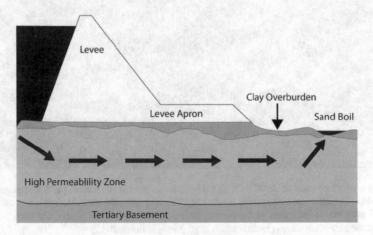

Fig. 5 Possible model for sand boil formation at Francis Levee Site

et al. 2007; Ivanov et al. 2006; Song et al. 2005; Kilty et al. 1986; Liechty 2010). In addition to using and processing seismic refraction and electrical resistivity methods separately, these methods can be used to simultaneously model the geophysical properties of the subsurface with the use of joint inversions. By using joint inversion, information obtained from one geophysical data set is used to constrain the inversion on another data set (Doetsch et al. 2012; Caterina et al. 2014; Zhou et al. 2014; Rittgers et al. 2015, 2016).

Cross-plot analysis has been extensively used in the oil industry for formation evaluations and lithology delineations since the 1960s (Fertl 1981; Krief et al. 1990; Shahin et al. 2009; Liu and Ghosh 2016; Anyiam et al. 2017; Holmes et al. 2017).

Cross-plot analysis using multiple input parameters such as electrical resistivity, nuclear, and acoustic logging have been used in determining lithological reservoir characteristics.

Cross-plot analysis is also used to combine and interpret SRT and ERT results on dams and levees. Cross-plot analysis using seismic refraction and electrical resistivity results is an approach where different soil types and conditions are classified based on their seismic velocity and electrical resistivity values (Hayashi and Konishi 2010; Inazaki and Hayashi 2011; Imamura et al. 2007). Instead of analyzing the results from each method separately, four quadrant criteria based on the ranges of seismic velocity and electrical resistivity are used. Based on the measured seismic velocity and resistivity at a given location, and where that point falls within the four quadrants, the soil type and integrity of dam or levee at that location is estimated. Therefore, in order to use cross-plot analysis, seismic velocity and electrical resistivity values that bound the four quadrants have to be determined.

One of the challenges of cross-plot analysis is determining the seismic velocity and electrical resistivity values that bound the four quadrants. These boundaries are determined by identifying common geotechnical traits to compromised zones, such as seepage and piping zones based on the geophysical methods used. Therefore, in the absence of geotechnical laboratory or borehole data that can be used to correlate geophysical and geotechnical properties of the soil, determining the bounds is difficult.

The objective of this study is to implement cross-plot analysis using two geophysical methods, seismic refraction tomography (SRT) and electrical resistivity tomography (ERT), in order to identify locations of preferential flow paths through the subsurface of the Francis levee that might have led to the formations of the three sand boils. A method of using theoretical modes, Archie's first law and effective fluid velocity model, to determine the seismic velocity and electrical resistivity bounds of the cross-plot is presented.

Methods

Seismic refraction surveying is one of the most commonly used seismic methods for engineering investigations (USACE 1995). Seismic refraction surveying is used to map the subsurface from recorded data using mechanical vibrations. Data is obtained by generating seismic energy at the surface at a known time and recording refracted energy using an array of geophones planted on the surface (Brosten et al. 2005).

Several factors affect seismic velocities through soils and rocks. Some of them include lithological properties of soils (grain sizes, grain shape, grain type, grain size distribution, amount of compaction, amount of consolidation and cementation), physical properties of soils (porosity, permeability, density, degree of saturation, pressure, and temperature), and elastic properties of soils [shear modulus (G), bulk modulus (K), Young modulus (E), Poisson's ratio (ν) and Lamé constant (λ)]. All the above factors are interrelated and affect the seismic velocity. For example, higher compaction will increase the shear and bulk modulus of the soil, reduce porosity, and therefore increase the seismic velocity through that material (Uyank 2011).

Electrical resistivity surveys provide continuous monitoring (in space) of the subsurface of dams and levees. Electrical resistivity surveying have been extensively used in the past to identify subsurface seepage through the body and foundation of dams (Butler et al. 1990; Okko et al. 1994; Abuzeid 1994; Panthulu et al. 2001; Zhou and Dahlin 2003; Kim et al. 2004; Lim et al. 2004; Cho and Yeom 2007; Sjödahl et al. 2008; Lin et al. 2013; Al-Fares 2014; Himi et al. 2016). Electrical resistivity of soil is mainly affected by porosity, degree of saturation, pore fluid resistivity, and clay content (Friedman 2005). Archie (1942) used clean sandstones and carbonates which were fully saturated with aqueous solutions of varying concentration and derived his first law, which states that the bulk resistivity (ρ_b) of a rock fully saturated with an aqueous fluid of resistivity (ρ_w) is directly proportional to the resistivity of the fluid. In his second law, Archie derived an empirical relation for partially saturated clean sand. Archie's first and second law do not incorporate the effect of clay on electrical resistivity. When clay is present, the path of the current is not just through the pores, but also along the surface of the clay material. Therefore, the measured bulk resistivity is now dependent on the clay content as well as the type of clay in the soil. Waxman and Smits (1968) derived a formula for calculating the bulk resistivity of soils containing clay.

Cross-plot criteria based on seismic and electrical attributes for seepage and piping are shown in Fig. 6. In the quadrant labeled 1, with low p-wave velocity and high resistivity, the dam is classified to be in poor condition. This condition is because low p-wave velocity is associated with poor compaction (low stiffness) and high porosity whereas high resistivity is associated with low clay content or a high porosity unsaturated sand. Therefore, a combination of these factors would classify a dam as a poor condition. Criteria for a good dam condition are shown in quadrant 4 with high p-wave velocity and low resistivity. A combination of well-compacted soil with low porosity (high p-wave velocity) and high clay content (low resistivity) is considered a good dam condition. Although low resistivity can also be associated with high porosity brine saturated sand, this is not a probable condition in dams and levees.

Study Area and Survey Parameters

Three locations were selected to conduct both seismic and electrical resistivity measurements. Survey line 1 is on the waterside of the levee, survey line 2 is on the berm of the levee, and survey line 3 is on the landside of the levee between the first sand boil and the two sets of sand boils. Each survey line is 478 m long and starts at the northern end and progresses southward parallel to the levee. Figure 7 shows the location of the three survey lines and the sand boils.

For the p-wave seismic refraction surveys, the whole length of each survey line is covered using a 24-geophone roll-along. Shot records were collected at 1 m offset from the first and last geophones and in-between all geophones. For electrical resis-

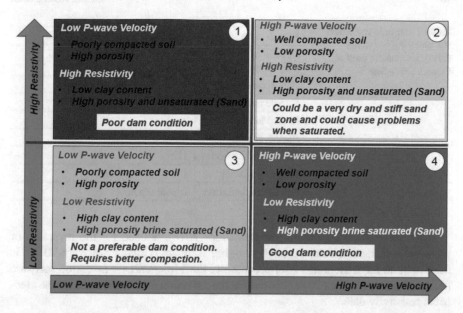

Fig. 6 Cross-plot analysis based on p-wave seismic velocity and electrical resistivity values. A compromised dam condition is characterized by a combination of low p-wave velocity and high resistivity whereas a good dam has high p-wave velocity and low resistivity

tivity surveys, the whole length of each survey line is covered using a 56-electrode roll along. Additional survey parameters are summarized in Table 1.

Fig. 7 Locations of p-wave seismic refraction and electrical resistivity survey lines. The arrows on the lines indicate direction of surveys

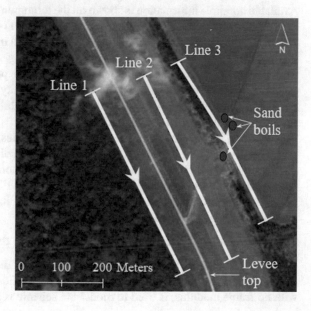

Table 1 Seismic refraction
and electrical resistivity
survey parameters used at
Francis Levee Site

Seismic refraction		
Number of geophones	48	10 Hz vertical component
Geophone spacing	2	m
Sample interval	0.125	ms
Record length	2	s
Electrical resistivity		
Number of electrodes	112	
Electrode spacing	1	m
Survey configuration	Dipole-dipole	

Rayfract™ software (Intelligent Resources 2018), is used for the inversion of all seismic refraction data. Surfer™ imaging software (Golden Software 2018), is used to build the tomograms after processing. EarthImager2D™ inversion software (Advanced Geosciences 2018) is used for the inversion and imaging of all the electrical resistivity data.

Results and Analysis

i. Waterside (Line 1)

The electrical resistivity tomograms for line 1 (waterside) are shown in Fig. 8. Available borehole information is shown on the tomograms to aid with interpretation. Based on borehole data, all eleven anomalies (E1–E11) on Fig. 8 are located in the sand zone. Ground water level (GWL) is at 4.7 m depth indicating that the sand zone is saturated.

A possible seepage zone has high permeability. The permeability of soil depends on porosity and grain size distribution. High permeability corresponds to high porosity, which leads to low p-wave velocity. High permeability also implies coarse soil or low clay content which leads to high resistivity.

Therefore, in order to reduce the number of anomalies in the ERT tomograms, the next step is to identify locations of low-velocity anomalies in the seismic refraction tomograms that are collocated with high resistivity anomalies. The p-wave seismic velocity tomograms for line 1 (waterside) are shown in Fig. 9. Six seismic anomalies (S1–S6), all located in the sand zone, are identified having low p-wave velocity compared to their background.

ii. Identification of true compromised zones using theoretical models

In order to identify true compromised zones out of the six possibilities shown in Fig. 9, an effective fluid velocity model (Eq. 1a) and (Eq. 1b), which is a suspension with no frame modulus, is used to model the seismic velocity of the sand zone and

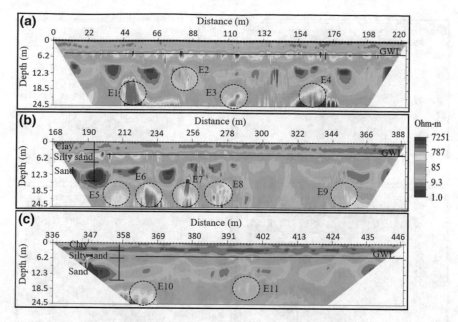

Fig. 8 Line 1 (waterside) electrical resistivity tomogram, **a** 0–220 m distance, **b** 168–388 m distance, and **c** 336–446 m distance. The broken line circles indicate locations of electrical resistivity anomalies (E1–E9) with high resistivity all located in the sand zone. GWL represents ground water level from well readings

to calculate porosity ($\phi_{velocity}$) from p-wave velocity tomograms.

$$K_{sat} = \left(V_p^2\right)\left[(1 - \phi_{velocity})\rho_o + (\phi_{velocity})\rho_w\right] \tag{1a}$$

$$K_{sat} = \frac{(K_s \cdot K_w)}{(1 - \phi_{velocity})K_w + (\phi_{velocity})K_s} \tag{1b}$$

where K_{sat} is the saturated bulk modulus of the soil, K_s is the sand grain bulk modulus (36.6 GPa), K_w is the bulk modulus of water (2.20 GPa), V_p is p-wave velocity (m/s), ρ_w is density of water (1000 kg/m^3), ρ_s is the sand grain density (2650 kg/m^3), and $\phi_{velocity}$ is porosity calculated from p-wave velocity. Since average p-wave velocity (V_p) can be obtained from the tomograms in Fig. 9, porosity ($\phi_{velocity}$) for each of the six seismic anomalies can be calculated by equating Eqs. (1a) and (1b).

Archie's first law (Eq. 2) for a fully saturated clean sand,

$$\rho_o = \rho_w \cdot \phi_{resistivity}^{-m}, \tag{2}$$

is used to model the electrical resistivity of a fully saturated clean sand and to calculate porosity ($\phi_{resistivity}$) where ρ_o is the bulk resistivity, ρ_w is the resistivity of the pore

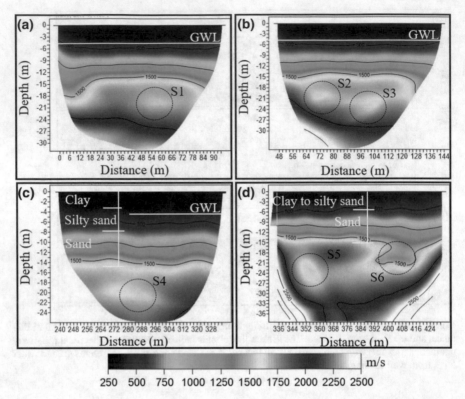

Fig. 9 Line 1 (waterside) p-wave tomogram, **a** 0–96 m distance, **b** 48–144 m distance, **c** 240–336 m distance, and **d** 336–432 m. The broken line circles indicate locations of seismic anomalies (S1–S6) with low p-wave velocity all located in the sand zone. GWL represents ground water level from well readings

fluid, $\phi_{resistivity}$ is porosity calculated from electrical resistivity, and cementation factor (m) is taken as 1.8 for sand.

From the seismic refraction tomogram (Fig. 9) and electrical resistivity tomograms (Fig. 8), the sand zone (excluding the anomalies) has an average p-wave velocity of 1750 m/s and an average electrical resistivity of 415 Ω m. Porosity of the sand zone is determined by using Eq. (1) and the average p-wave velocity of the sand zone. The average resistivity of the sand zone is then used to determine the resistivity of the pore fluid (ρ_w) using Eq. (2). Resistivity of the pore fluid (ρ_w) is assumed to remain constant and porosity ($\phi_{resistivity}$) for each of the six seismic anomalies can be calculated using Eq. (2). Once the porosities of the six anomalies are calculated using the two models, their consistency is compared as shown in Table 2. Seismic anomaly 5 (S5), located 18–27 m below the surface, has both calculated porosities within an acceptable range (<0.5) and is the most consistent between the two models (Table 2). Therefore, seismic anomaly 5 (S5) is considered an area most probably associated with a seepage path.

Table 2 Average p-wave velocity and electrical resistivity values for the six anomalies on the waterside, summary of values and consistency in the calculated porosity of the six anomalies

Anomalies (Fig. 9)	Velocity (m/s)	Resistivity (Ω m)	$\phi_{velocity}$ (Archie's first law model)	$\phi_{resistivity}$ (effective fluid velocity model)	$\Delta\phi = \phi_{res} - \phi_{vel}$
S1	1656	160	0.35	0.49	0.14
S2	1597	133	0.40	0.54	0.14
S3	1597	63	0.40	0.82	0.42
S4	1646	15	0.35	1.82	1.47
S5	1639	218	0.36	0.41	0.05
S6	1430	29	0.71	1.26	0.55

iii. Landside (Line 3)

The electrical resistivity tomograms for line 3 (landside) are shown in Fig. 10. The survey for Fig. 10a was conducted in May whereas the survey for Fig. 10b, c were conducted in December. Anomaly EA2 in Fig. 10b is not present in Fig. 10a,

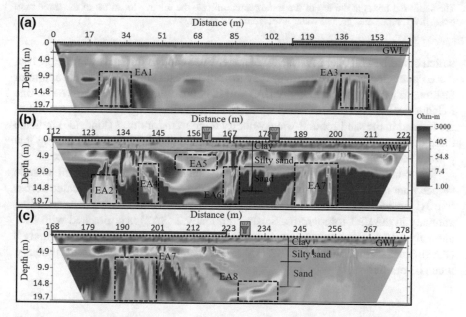

Fig. 10 Electrical resistivity tomograms for line 3 (landside). **a** 0–170 m distance, **b** 112–222 m distance, and **c** 168–278 m distance. The broken line boxes indicate locations of electrical resistivity anomalies (EA1–EA8) with high resistivity all located in the sand zone. The red and blue lines at the top of the figures indicate areas of overlap. The small red boxes at the top of the tomograms indicate the relative locations of the three sand boils. GWL represents ground water level from well readings

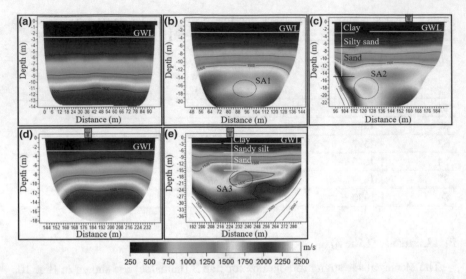

Fig. 11 Line 3 (landside) p-wave velocity tomograms. **a** 0–96 m distance, **b** 48–144 m distance, **c** 96–192 m distance, **d** 144–240 m distance, and **e** 192–288 m distance. The broken circles indicate locations of seismic anomalies (SA1–SA3) with low p-wave velocity all located in the sand zone. The small red boxes at the top of the tomograms indicate the relative locations of the three sand boils. GWL represents ground water level from well readings

which could be an effect of seasonal change. Even though the surveys for Fig. 10b, c are conducted at the same time, they do not indicate anomaly EA7 in a similar fashion. In general, since anomaly EA3 and EA4 are collocated, ERT survey on line 3 (landside) indicate seven distinct anomalies where all the high resistivity anomalies are located in the sand zone. P-wave velocity tomogram for line 3 (landside) indicate three distinct seismic anomalies labeled SA1, SA2, and SA3 (Fig. 11) having low p-wave velocity compared to their background and all located in the sand zone.

The location of all the electrical resistivity anomalies (Fig. 10) and all seismic velocity anomalies (Fig. 11) are obtained from the tomograms and indicated across line 3 (landside) as shown in Fig. 12. Seismic anomalies 2 and 3 have a corresponding anomaly in the ERT tomograms. Seismic anomaly 2 (SA2) is collocated with ERT anomaly 3 (EA3) and seismic anomaly 3 (SA3) is collocated with ERT anomaly 8 (EA8). Since seismic anomaly 1 (SA1) has no corresponding high resistivity anomaly, it can be omitted.

Cross-Plot Analysis

In order to identify which of the remaining two pairs of collocated anomalies, ([SA2 and EA3] and [SA3 and EA8]), is associated with the formation of the sand boils, a cross-plot analysis based on the seismic velocity and electrical resistivity of anomaly

Fig. 12 An aerial image of survey line 3 (landside) with locations of SRT and ERT anomalies indicated along the survey line

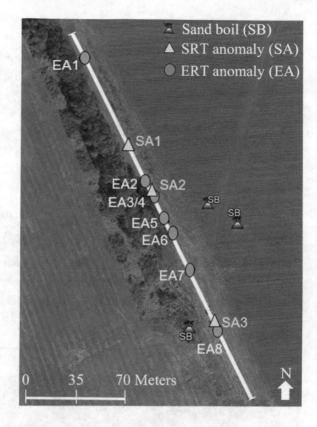

number 5 (S5) on the waterside is performed. Cross-plot analysis using the seismic velocity (1639 m/s) and electrical resistivity (218 Ω m) values of anomaly number 5 (S5) (Table 2) on the waterside can be used to identify similar anomalous locations on the landside (line 3). This is a variant of standard cross-plot analysis where prior information is used as the boundary of the cross-plot analysis.

The cross-plot analysis of seismic anomaly 2 (SA2) and electrical anomaly 3 (EA3) is shown in Fig. 13. The blue boxes on both the ERT (Fig. 13a) and SRT (Fig. 13b) is a background area selected covering both anomalies (SA2 and EA3). The size (area) of the blue boxes are the same on both ERT and SRT tomograms. Each grid point in the ERT tomogram has a corresponding grid point in the SRT tomogram. All electrical resistivity and p-wave velocity pairs within the blue boxes (background) are cross-plotted on a resistivity versus p-wave velocity plot and shown with blue dots (Fig. 13c). Similarly, the cross-plot for the SA2 (black points) and EA3 (purple points) are shown in Fig. 13c. The cross-plot boundary values of 1639 m/s and 218 Ω m are shown with the red lines on the cross-plot Fig. 13c. Analysis of the cross-plot indicates that neither SA2 nor EA3 fall in the compromised quadrant.

Similarly, the cross-plot analysis for SA3 and EA8 is shown in Fig. 14c. A smaller area shown with a white box on both tomograms is picked by focusing on the high

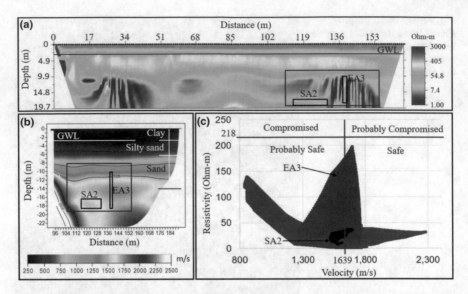

Fig. 13 **a** Electrical resistivity tomograms for line 3 (landside) for 0–170 m distance, **b** line 3 (landside) p-wave velocity tomograms for 96–192 m distance, **c** cross-plot analysis using SA2 and EA3

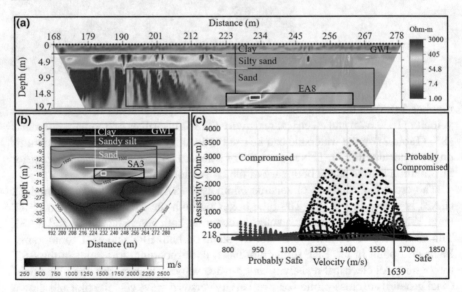

Fig. 14 **a** Electrical resistivity tomograms for line 3 (landside) for 168–278 m distance, **b** line 3 (landside) p-wave velocity tomograms for 168–278 m distance, **c** cross-plot analysis using SA3 and EA8

resistivity area within SA3 (black box). The cross-plot of this smaller area is shown by the light blue dots on the cross-plot (Fig. 14c). The cross-plot analysis shows that the majority of SA3/EA8 falls within the compromised quadrant. This is an indication that the combination of SA3 and EA8 have similar features to anomaly number 5 (S5) on the waterside (Fig. 9d) and could be associated with the formation of the three sand boils.

Interpretation

In order to identify a possible seepage path, the location of the sand boils, the six anomalies on the waterside, and the three anomalies on the landside are plotted on an aerial image of Francis Levee Site as shown in Fig. 15. A possible seepage line parallel to the northern edge the meander belt is drawn passing through anomaly S5 on the waterside and anomaly SA3/EA8 on the landside. The projection of this possible seepage path passes through the three sand boils. Flow path parallel to the meander is expected because the soil deposit inside the meander has low compaction and high permeability compared to the native ground. Water can flow through the highly permeable sand and gravely sand and cause sand boil formations at locations where the overburden clay layer is thin. The anomaly on the landside is located 18–27 m below the surface whereas on the waterside the anomaly is between 15 and 21 m below the surface.

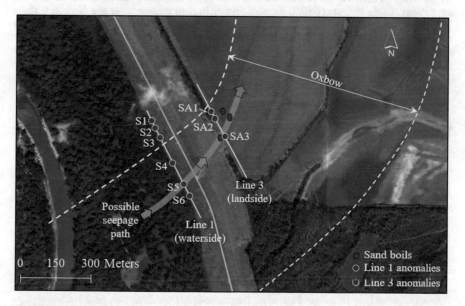

Fig. 15 Possible seepage path (blue line) going parallel to the northern edge the meander belt and passing through anomaly S5 on the waterside and anomaly SA3/EA8 on the landside

Conclusions

Results from seismic refraction and electrical resistivity surveys conducted at Francis Levee Site indicate seven distinct anomalies that might be associated with seepage. For both surveys on the waterside and landside, applying the restriction that velocity of a seepage zone should be lower than the background velocity reduces the number of anomalies in the ERT tomograms.

The location (depth from the surface) of the anomalies, associated with the proposed seepage paths, support the idea that the preferential flow occurs within the sand layer in the old oxbow. The location of the sand boils is along the proposed trajectory but most likely outcrops where the overlying impermeable clay layer is thin or the weakest. The predicted subsurface pathway is reasonable because it follows the contour of the meander and it passes through the location of the sand boils.

Multiple geophysical surveys such as seismic refraction tomography and electrical resistivity surveys can be used for the early identification of compromised zones in dams and levees. Instead of analyzing the results from multiple methods separately, cross-plot analysis can be used to combine the strength of the individual methods and provide a simplified representation of the integrity of the dam or levee.

Acknowledgements Funding for this report was provided by the NSF and Colorado School of Mines, Partnership for International Research Development (PIRE) program, Award #OISE-1243539/400512 and the United States Department of Agriculture (USDA). We would like to thank the U.S. Army Corps of Engineers for equipment loan, Peter Nimrod (Mississippi Levee Board) for providing us access to the site, Joe Dunbar (ERDC) for discussion and input into the underlying geology, Justin Rittgers (USBR) for participating in field measurement and discussion about geophysical interpretation, Wim Kanning and Carolyne Bocovich (Colorado School of Mines) for participating in field measurements and discussion of model of failures, Trent Snellings (USGS) for participating in field measurement, and colleagues at the National Center for Physical Acoustics (NCPA) that participated in data collection.

References

Abuzeid, N., (1994), "Investigation of channel seepage areas at the existing Kaffrein dam site (Jordan) using electrical-resistivity measurements," *Journal of Applied Geophysics*, 32, pp. 163–175.

Advanced Geosciences Inc., (AGI), 2018. http://www.agiusa.com/index.shtml.

Al-Fares, W., (2014) "Application of Electrical Resistivity Tomography Technique for Characterizing Leakage Problem in Abu Baara Earth Dam, Syria," *International Journal of Geophysics*, 2014, Article ID 368128.

Anyiam, O. A., Mode, A. W., Okara, E. S., (2017), "The use of cross-plots in lithology delineation and petrophysical evaluation of some wells in the western Coastal Swamp, Niger Delta," Journal of Petroleum Exploration and Production Technology, 8, pp. 61–71.

Archie, G.E., (1942), "The electrical resistivity log as an aid in determining some reservoir characteristics," *Society of Petroleum Engineers, Transactions of the AIME*, 146, pp. 54–62.

Brackett, T. C., (2012), Use of Geophysical Methods to Map Subsurface Features at Levee Seepage Locations, Master's thesis, University of Mississippi.

Brosten, T. R., Llopis, J. L., and Kelley, J. R., (2005) "Using geophysics to assess the condition of small embankment dams," *USACE Engineering Research and Development Center*, ERDC/GSL TR-05-17.

Butler, D. K., Llopis, J. L., Dobecki, T. L., Wilt, M. J., Corwin, R. F., Olhoeft, G., (1990), "Comprehensive geophysical investigation of an existing dam foundation," *The Leading Edge*, 9, pp. 44–53.

Caterina, D., Hermans, T. & Nguyen, F., (2014) "Case studies of prior information in electrical resistivity tomography: comparison of different approaches," *Near Surface Geophysics*, 12(4), pp. 451–465.

Cho, I., and Yeom, J., (2007), "Crossline resistivity tomography for the delineation of anomalous seepage pathways in an embankment dam," *Geophysics*, 72, pp. G31–G38.

Dahlin, T., and Johansson, S., (1995), "Resistivity variations in an earth embankment dam in Sweden," *1st EEGS Meeting, Environmental Risk and Planning*.

Doetsch, J., Linde, N., Pessognelli, M., Green, A.G. & Gunther, T., (2012) "Constraining 3-D electrical resistance tomography with GPR reflection data for improved aquifer characterization," *Journal of Applied Geophysics*, 78, pp. 68–76.

Golden software (2018), http://www.goldensoftware.com/products/surfer.

Google Earth 2013, © 2013 Google Inc., https://www.google.com/earth/.

Google Earth 2015, © 2015 Google Inc., https://www.google.com/earth/.

Fertl, W. H., (1981), "Openhole Cross-plot Concepts – A Powerful Technique in Well Log Analysis," Journal of Petroleum Technology, 33, https://doi.org/10.2118/8115-pa.

Fisk, H.N., (1944), Geological Investigation of the Alluvial Valley of the Lower Mississippi River: Mississippi River Commission, Vicksburg, MS., Plate 22.

Friedman, S.P., (2005), "Soil properties influencing apparent electrical conductivity: a review," *Computers and Electronics in Agriculture*, 46 (1-3), pp. 45-70.

Hayashi, K., and Konishi, C., (2010), "Joint use of a surface-wave method and a resistivity method for safety assessment of levee systems," *ASCE, GeoFlorida 2010*, pp. 1340–1349.

Hickey, C. J., Eikmov, A., Hanson, G. J., and Sabatier, J.M., (2009), "Time lapse seismic measurements on a small earthen embankment during an internal erosion experiment," *Symposium on Applications of Geophysics to Environmental and Engineering Problems (SAGEEP)*, Fort Worth, TX, pp. 144–156.

Hickey, C. J., Mathias, J. M., Robert, R. W., and Wodajo, L. W., (2015), Geophysical methods for the assessment of earthen dams, *Advances in Water Resources Engineering*, Springer, USA.

Holmes, M., Holmes, A. M., & Holmes, D. I., (2017), "Mixed Reservoir Wetting in Unconventional Reservoirs and Interpretation of Porosity/Resistivity Cross Plots, Derived From Triple-Combo Log Data," Society of Petrophysicists and Well-Log Analysts (SPWLA) 58th Annual Logging Symposium, Oklahoma City, OK, June 17–21, 2017.

Imamura, S., Tokumaru, T., Mitsuhata, Y., Hayashi, K., and Inazaki, T., (2007) "Application of integrated geophysical techniques to vulnerability assessment of levee," *Proceeding of the 116th SEGJ Conference*, pp. 120–124.

Himi, M., Casado, I., Sendros, A., Lovera, R., Casas, A., and Rivero, L., (2016) "Using the Resistivity Method for Leakage Detection at Sant Llorenç de Montgai Embankment (Lleida, NE Spain)," *Near Surface Geoscience 2016 - 22nd European Meeting of Environmental and Engineering Geophysics* , Barcelona, Spain.

Inazaki, T., and Hayashi, K., (2011), "Utilization of integrated geophysical surveying for the safety assessment of levee systems," *Symposium on Applications of Geophysics to Environmental and Engineering Problems (SAGEEP)*, Charleston, South Carolina.

Intelligent resources INC (2018). http://www.rayfract.com/.

Ivanov, J., Miller, R. D., Stimac, N., Ballard, R. F., Dunbar, J.B., and Smullen, S., (2006), "Time-lapse seismic study of levees in southern New Mexico," SEG expanded abstract.

Kilty, K. T., Norris, R. A., McLamore, W. R., Hennon, K. P., and Euge, K., (1986), "Seismic refraction at Horse Mesa Dam: an application of the generalized reciprocal method," *Geophysics*, 51, pp. 266–275.

Kim, H. S., Oh. S. H., Park, H. G., Chung, H. J., Shon, H. W., Kim, K. S., Lim, H. D., (2004), "Electrical resistivity imaging for comprehensive evaluation of earth-rock dam seepage and safety," *Procs. International Commission on Large Dams (ICOLD)*, 72th Annual Meeting, Seoul, Korea, pp. 123–134.

Kim, J. H., Yi, M. J., Song, Y., Seol, S. J., Kim, K. S., (2007), "Application of geophysical methods to the safety analysis of an earth dam," *Journal of Environmental Engineering Geophysics*, 12, pp. 221–235.

Krief, M., Garat, J., Stellingwerff, J., and Ventre, J., (1990), "A Petrophysical Interpretation Using the Velocities of P and S Waves (Full-Waveform Sonic)," Society of Petrophysicists and Well-Log Analysts, 1990, ID: SPWLA-1990-v31.

Liechty, D. J., (2010), "Geophysical surveys, levee certification geophysical investigations, DC resistivity," *23rd EEGS Symposium on the Application of Geophysics to Engineering and Environmental Problems*.

Lim, H. D., Kim, K. S., Kim, J. H., Kwon, H. S., Oh, B. H., (2004), "Leakage detection of earth dam using geophysical methods" *Procs. International Commission on Large Dams (ICOLD)*, 72th Annual Meeting, Seoul, Korea, pp. 212–224.

Lin, C. P., Hung, Y. C., Yu, Z. H., and Wu, P. L., (2013), "Investigation of abnormal seepages in an earth dam using resistivity tomography," *Journal of GeoEngineering*, 8, pp. 61–70.

Liu, C., and Ghosh, D. P., (2016), "AVO and Spectral Crossplot: A Case Study in the Malay Basin," Offshore Technology Conference, https://doi.org/10.4043/26482-ms.

Nimrod, P., (2011). 2011 Flood Report, A Success Story, Mississippi Levee Board.

Okko, O., Hassinen, P., Korkealaakso, J., (1994), "Location of leakage paths below earth dams by geophysical techniques," *Proceedings International Conference on Soil Mechanics and Foundation Engineering (ICSMFE)*, New Delhi, pp. 1349–1352.

Panthulu, T. V., Krishnaiah, C., Shirke, J. M., (2001), "Detection of seepage paths in earth dams using self-potential and electrical resistivity methods," *Engineering Geology*, 59, pp. 281–295.

Rittgers, J.B., Revil, A., Planes, T., Mooney, M.A. & Koelewijn, A.R., (2015) "4-D imaging of seepage in earthen embankments with time-lapse inversion of self-potential data constrained by acoustic emissions localization," *Geophysics Journal International*, 200, pp. 758–772.

Rittgers, J.B., Revil, A., Mooney, M.A., Karaoulis, M., Wodajo, L., and Hickey, C., (2016) "Time-lapse joint inversion of geophysical data with automatic joint constraints and dynamic attributes," *Geophysics Journal International*, 207, pp. 1401–1419.

Saucier, R. T., (1994), Geomorphology and quaternary geologic history of the lower Mississippi valley, U.S. Army Engineer Waterways Experiment Station, Volume 1.

Shahin, A., Stoffa, P. L., Tatham, R. H., and Sava, D., (2009), "Multicomponent Seismic Time-lapse Cross-plot And Its Applications," Society of Exploration Geophysicists (SEG) International Exposition and Annual Meeting, Houston, TX.

Sjödahl, P., Dahlin, T., Johansson, S., and Loke, M. H., (2008), "Resistivity monitoring for leakage and internal erosion detection at Hällby embankment dam," *Journal of Applied Geophysics*, 65, pp. 155–164.

Song, S. H., Song, Y., and Kwon, B. D., (2005), "Application of hydrogeological and geophysical methods to delineate leakage pathways in an earth fill dam," *Exploration Geophysics*, 36, pp. 92–96.

US Army Corps of Engineers (USACE) EM 1110-1-1802 (1995) *Geophysical exploration for engineering and environmental investigations*, Engineer Manual.

Uyank, O., (2011), "The porosity of saturated shallow sediments from seismic compressional and shear wave velocities," *Journal of Geophysics*, 73, pp. 16–24.

Waxman, M.H., and Smits, L.J.M., (1968), "Electrical Conductivities in Oil-Bearing Shaly Sands," *Society of Petroleum Engineers Journal*, 8, pp. 107–122.

Zhou, B., and Dahlin, T., (2003), "Properties and effects of measurement errors on 2D resistivity imaging surveying," *Near Surface Geophysics*, 1, pp. 105–114.

Zhou, J., Revil, A., Karaoulis, M., Hale, D., Doetsch, J. & Cuttler, S., (2014) "Image-guided inversion of electrical resistivity data," *Geophysics Journal International*, 197, pp. 292–309.

A Borehole Seismic Reflection Survey in Support of Seepage Surveillance at the Abutment of a Large Embankment Dam

Karl E. Butler, D. Bruce McLean, Calin Cosma and Nicoleta Enescu

Abstract Retrofitting existing dams for the installation of modern monitoring instrumentation requires confidence in one's knowledge of the dam's internal structures. In 2010, the operator of the Mactaquac Generating Station wished to install a fibre optic distributed temperature sensing (DTS) cable as close as possible to the sub-vertical contact between the concrete diversion sluiceway and the clay till core of the adjacent zoned embankment dam. Given a lack of detailed as-built drawings, a plan was developed to image the interface by GPR or seismic reflection surveying from a sub-parallel borehole, offset by approximately 1 m at surface and by an estimated 4 m at the dam's foundation, near 50 m depth. Seismic reflection imaging, although novel for this application, emerged as the favoured approach after the range of borehole GPR surveys proved inadequate, due to high electrical conductivity in the concrete. A very high resolution wall-clamping seismic tool, with piezoelectric source and eight receivers, was operated in the dry borehole at 60 cm increments. Two surveys with different tool orientations were conducted to favour the reception of either P-wave or S-wave reflections from the interface, although such reflections were obscured, in the shot records, by relatively slow surface waves travelling along the borehole wall. A relatively simple processing flow involving bandpass filtering, CMP (common midpoint) stacking, and mean trace subtraction was successful in revealing an interpreted S-wave reflection from the interface, having a dominant frequency near 7 kHz representing a wavelength of about 35 cm. Interference from residual surface waves and apparent scattering from concrete layers, rebar or other heterogeneities near the borehole was very significant, but near-agreement of the interpreted reflection with the interface shown on engineering design plans provided confidence to proceed with the installation of two monitoring boreholes estimated to lie with 50 cm of concrete/clay contact.

K. E. Butler (✉)
Department of Earth Sciences, University of New Brunswick,
Fredericton, NB E3B 5A3, Canada
e-mail: kbutler@unb.ca

D. B. McLean
Mactaquac Generating Station, NB Power, Keswick Ridge, NB E6L 1B2, Canada

C. Cosma · N. Enescu
Vibrometric Canada Limited, Toronto, ON, Canada

© Springer Nature Switzerland AG 2019
J. Lorenzo and W. Doll (eds.), *Levees and Dams*,
https://doi.org/10.1007/978-3-030-27367-5_3

41

Introduction

As the average age of dams and levees worldwide continues to increase, and original design lives are approached or exceeded, the need for non-invasive to minimally-invasive methods of investigation and monitoring has grown (e.g., FEMA 2005, 2015; FERC 2017; ASCE 2017; CEATI 2018). Geophysical methods provide a means of investigating the internal structure of dams but factors such as size, steep topography, protective rip-rap, and partial cover by contained water or waste can make it challenging to deploy geophysical methods on surface and achieve sufficient resolution at required depths of exploration. In some situations, borehole geophysical surveys can yield much better results, particularly where the target zone is small compared to its depth, or is a sub-vertical internal feature such as a low-permeability core or an abutment.

This investigation focuses on one such target at the Mactaquac Generating Station—a 660 MW hydroelectric facility located on the Saint John River, approximately 20 km upriver of Fredericton, New Brunswick, Canada (Fig. 1). The facility, completed in 1968, includes a 500 m long zoned rockfill embankment (Conlon and Ganong 1966; Tawil and Harriman 2001), which stands up to 58 m high above its foundation or approximately 32 m above its toe (Fig. 2). In 2010, the dam's operator NB Power, had purchased a fibre optic distributed temperature sensing (DTS) system to be installed in such a way as to monitor for any indications of preferential seepage along the ~50 m long sub-vertical interface between the embankment

Fig. 1 Location of the Mactaquac hydroelectric generating station (yellow circle) in the province of New Brunswick (shaded red), Canada

Fig. 2 Aerial photo of the Mactaquac hydroelectric generating station, with its 500 m long zoned rockfill embankment dam

dam's compacted clay till core and the adjacent concrete diversion sluiceway structure (Figs. 2 and 3). The idea was to install the fibre optic cable in a borehole drilled into the concrete as close as possible to the interface, so as to maximize the DTS system's sensitivity to any temperature anomalies caused by interfacial seepage flow. Given the fact that as-built drawings were not available, the first step in this process was to determine the precise position of the concrete/clay core interface. The methods selected for interface delineation were single-borehole GPR (ground penetrating

Fig. 3 Close-up view of the interface between the embankment dam and the steeply inclined concrete South End Pier (SEP) of the Diversion Sluiceway. Inset shows a 3D model of the interface with the embankment removed

radar) reflection surveying as described by Giroux et al. (2011), and single-borehole seismic reflection surveying—the topic of this paper.

NB Power's interest in investigating conditions along the northern abutment of the embankment dam was motivated by the fact that concrete structures at the station suffer from expansion and degradation as a consequence of Alkali Aggregate Reactivity (AAR). Innovative techniques have been developed onsite for ongoing remediation of the concrete structures (e.g. Gilks et al. 2001), and it was considered important to be proactive by initiating an investigation of the interface between the embankment dam and concrete diversion structure. Apart from the AAR issue, interfaces between earthfill/rockfill embankments and concrete structures are generally recognized as areas of elevated risk for the development of seepage, internal erosion and damaging piping within embankment dams (Fell et al. 2005; Mattsson et al. 2008). Seepage can arise if a gap opens along the interface as a consequence of gradual or earthquake-induced settlement of the fill, or differential thermal expansion and contraction.

Survey Design

The geometry of the problem, and the idea of imaging the interface from a single borehole using reflected seismic waves (P-waves and S-waves) is illustrated in Fig. 4. The objective was to delineate the interface with decimetre precision, so that a borehole for the DTS cable could be installed in the concrete as close to the embankment dam core as possible. A surveying borehole (SEPI-1), approximately 50 m in length and having a diameter of 3.782″ (96.06 mm), was drilled into the concrete South End Pier (SEP) of the diversion sluiceway during the summer of 2010 from a point on the road crossing the structure (Figs. 3 and 4). As a further aid to visualization, Fig. 5 shows the approximate location of this borehole when superimposed, schematically, on a photograph taken during construction of the concrete SEP in 1965. The borehole was continuously cored, yielding concrete core samples with porosities of 11–16%, containing abundant angular fragments of greywacke aggregate (Fig. 6).

As shown in Fig. 4, the surveying hole was drilled a safe distance back from the interface, and at a steeper angle than the expected dip of the interface so that interface reflections could be more readily separated (during data processing) from refracted, or guided waves travelling directly from source to receiver along the borehole itself. Based on dam design plans and on a survey of the borehole inclination and collar location, the interface was expected to lie 1 m away from the hole at surface and 4 m away at the bottom (Fig. 4). The range of distances was considered better than a constant borehole-interface separation for the afore-mentioned reason and because it was impossible to predict an optimal separation in advance. If the separation was too small then the available seismic systems would not have the resolution required to distinguish the reflected seismic arrivals from direct arrivals. On the other hand, if the separation was too large, the high frequency/short wavelength seismic systems needed for high resolution imaging would not have the required depth of penetration.

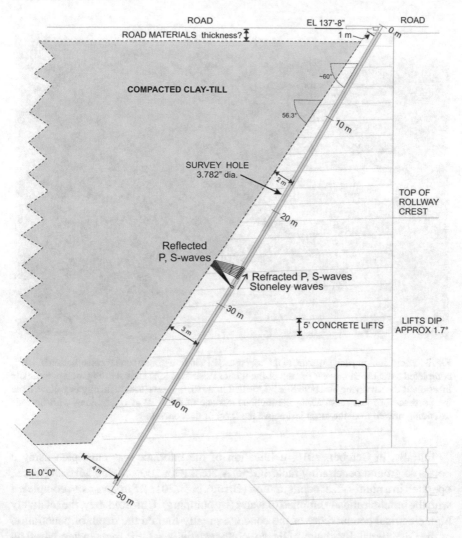

Fig. 4 Long-section AA′ (looking upstream) though the embankment/SEP interface showing the geometry used for multi-channel single hole seismic reflection imaging of the interface. Shape of the SEP has been traced from an engineering design plan. Internal sub-horizontal lines represent contacts between concrete pour lifts which were nominally 5′ (1.5 m) thick. Borehole-to-interface distances are estimates based on dam design plans and a borehole orientation survey

Fig. 5 Photo, looking downstream, of the concrete SEP (in foreground) under construction in 1965, completed to an elevation near the top of the tunnel (visible at left) shown in Fig. 4. Approximate location of the surveying borehole SEPI-1, drilled in 2010, is shown schematically in yellow. Note the presence of rebar mesh along the inclined surface of the SEP at right, along with tie rods extending inward from the mesh just above the level of the concrete

Initially, in October, 2010, delineation of the interface was attempted using a borehole ground penetrating radar (GPR) system with 100 and 250 MHz antennas, operated in a multi-offset configuration (Giroux et al. 2011). Surveys were completed with the borehole mostly emptied of water (by pumping). Unfortunately, the relatively high electrical conductivity of the concrete greatly limited the depth of penetration of the GPR signal. Estimates of the concrete resistivity at GPR frequencies, based on measurements of GPR velocity, attenuation, frequency-lowering were in the range of only 10–25 Ωm. (Later measurements of DC resistivity, by normal resistivity logging, ranged from 15 to 90 Ωm.) A possible interface was detected over a portion of the borehole but it was considered to be too close to the borehole and too steeply inclined to be valid.

Following the GPR survey, the possibility of seismic imaging was pursued. A review of the geophysical literature revealed that experimental surveys of high resolution single borehole seismic reflection imaging had been carried out by oil and gas industry for the detection of geological boundaries and fractures at distances on the order of 1–10 m from the borehole. These surveys (e.g. Hornby 1989; Emersoy et al. 1998; Chabot et al. 2002; Franco et al. 2006; Tang et al. 2007; Bing et al. 2011) had utilized modified full waveform sonic (FWS) logging tools with relatively powerful

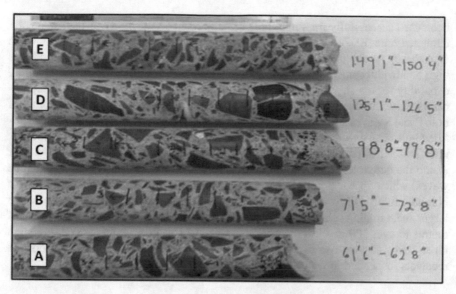

Fig. 6 Five concrete core samples (2.5″ or 63.5 mm diameter) collected between depths of 20 and 45 m in borehole SEPI-1. Aggregate pieces are greywacke. Fractional porosities determined by hydrostatic weighing of three 30 mm thick disks from each sample varied from 0.11 to 0.16 (±0.02)

sources, relatively long source-receiver offsets, and up to 24 receivers. The large number of channels allowed the use of advanced data processing techniques for the enhancement of reflections, relative to other interfering seismic arrivals. While those tools and services were not considered feasible for the relatively short and small borehole at Mactaquac, they did offer proof-of-concept that seismic imaging was feasible.

In late August 2012, two different high frequency borehole seismic systems were used for reflection surveying. The first was a 4-channel slim-hole FWS logging tool from Mount Sopris Instrument Company Ltd., modified by the use of both standard and custom-length sonic isolator sections, to allow the acquisition of data at 11 different source-receiver offsets. The FWS survey was considered speculative in that the instrument is designed for the measurement of seismic wave velocities in water-filled holes—not for the measurement of reflections from off-hole interfaces. There were however, logistical advantages in terms of instrument availability, rental cost, and ease of use. And even if the tool was not successful in imaging the interface, it would provide excellent control over P and S-wave velocity within the concrete, which would be valuable for purposes of processing data from the second seismic imaging survey, described below.

The second seismic reflection survey employed a unique, very high resolution, 8 channel borehole seismic system, known as the model PS-8R, originally developed by Vibrometric for research on rock mass characterization at hard rock nuclear waste repository sites (Cosma 1995; Emsley et al. 1997). The instrument had very recently been used successfully for single-hole seismic reflection imaging of the excavation damaged zone (EDZ) surrounding a tunnel at very short ranges of 0.8–1.4 m (Cosma et al. 2010). Although recognized to be experimental, this unique instrument with its high bandwidth sonic to ultrasonic source was considered to offer the best imaging capability for the short range application at Mactaquac dam. Perhaps most significantly, its ability to clamp directly to the borehole wall would promote coupling of both P and S waves into the concrete, and would allow the survey to be done in an air-filled hole. The direct wall coupling and absence of water was expected to eliminate interference from direct and trapped P-wave modes travelling from source to receiver inside the borehole, and also to reduce interference from Stoneley waves (or tube waves) which appear especially strong when measured using hydrophones inside a borehole filled with water. The survey would benefit from experience acquired in larger scale mining-related applications with target distances ranging from 10s of metres to over 100 m from the borehole (e.g., Cosma et al. 2006, 2007), although those surveys made use of more powerful piezoelectric sources and hydrophone receivers suspended in water.

Instrumentation and Data Acquisition

Full Waveform Sonic (FWS) Logging

The Mount Sopris FWS logging tool (model 2SAFF-1000, Fig. 7) consisted of a transmitter section equipped with an acoustic source separated from a receiver section by a flexible rubber hose known as a sonic isolator. The variable frequency source was operated in monopole mode with a centre frequency of 15 kHz which appeared to provide the strongest P and S-waves arriving by refraction along the borehole wall. The receiver section included four hydrophone receivers at offsets of 3, 4, 5, and 6 ft from the source when using the standard 2.5 ft sonic isolator.

Although the FWS logging tool was designed to measure only 'direct' waves, travelling from source to receiver along the borehole, special efforts were made to improve the chances of detecting weaker reflections from off-hole interfaces. In particular, additional source-receiver offsets of 3.5, 4.5, 5.5, 6.5, 7, 8, and 9 ft were acquired using a custom-made 3 ft-long sonic isolator, either on its own or in combination with the standard isolator on successive logging runs. The point of acquiring data at so many source-receiver offsets was to facilitate more effective use of multi-channel processing methods and common midpoint stacking for enhancing reflections from off-hole interfaces. Logging runs were also acquired very slowly (at about 0.6 m/min) to minimize motional noise and maximize signal averaging while maintaining a fine measurement interval (a stack of 16 pulses—the maximum available—being acquired each 2.49 cm depth).

Strong Stoneley wave arrivals (surface waves that propagate along the well bore at a velocity lower than that of the S-wave) are inevitable with conventional (non-

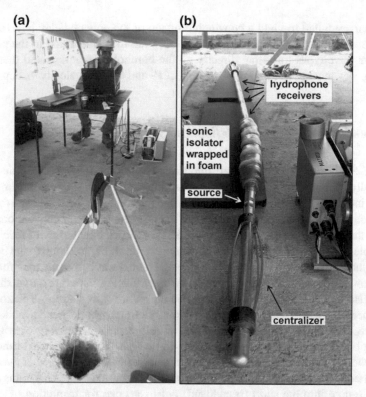

Fig. 7 **a** Surface components of the Mount Sopris FWS logging system, including winch with control console attached, shown while logging borehole SEPI-1. **b** FWS logging tool with closed cell foam wrapped around sonic isolator for Stoneley wave suppression

clamping) sonic logging tools that use acoustic sources and hydrophone receivers suspended in water-filled boreholes. Stoneley waves represented interference in this survey given that they dominated the recordings at times when much weaker reflections from the concrete/clay interface were expected to arrive. Various types of experimental Stoneley wave suppressors including closed cell foam (Milligan et al. 1997), rubber torque arrestors (intended to limit twisting of pumps in water wells), and a custom-machined plastic collar were attached to the sonic isolator in attempts to attenuate Stoneley wave propagation along the borehole by filling as much as the borehole diameter as possible. The very good (smooth) condition of the borehole wall allowed near-complete filling of the hole (while still allowing the tools to be easily lowered and raised easily). However, Stoneley waves still dominated the records during most of the time interval when reflections from the concrete/clay interface were expected to arrive. More sophisticated suppressors (e.g. Daley et al. 2003; Green wood et al. 2012), used successfully for lower frequency surveys, were not available. And attempts to attenuate Stoneley waves by velocity filtering of multi-offset shot records during processing were ineffective due to spatial aliasing. Ultimately, the

FWS survey failed to yield a convincing image of the concrete/clay interface; any P-wave interface reflections were simply too weak to be identified in the presence of other wave types travelling directly up the borehole. As anticipated then, FWS logging was most useful in providing logs of P- and S-wave velocity vs depth.

Single-Borehole Seismic Reflection Surveying

The PS-8R borehole tool is a high-frequency (sonic to ultrasonic) seismic system originally designed for characterizing conditions surrounding boreholes or tunnels drilled into rock. The nominal bandwidth of the instrument is 5–60 kHz and the range of investigation can exceed 20 m in rock. The PS8R is similar to a conventional FWS tool in that it includes a seismic source separated from a linear array of receivers by a flexible joint. However, it differs significantly in its ability to clamp to the borehole wall—thereby improving the radiation of both P and S waves into the rock formation and allowing it to be used in dry boreholes so that Stoneley waves are diminished. The PS8R also differs from FWS logging tools in the nature of its source—a piezoelectric transducer that emits pulse-trains of pseudo-random lengths at pseudo-random time intervals. The pulse trains are very weak (barely detectable audibly or with one's finger placed on the transducer) but signal-to-noise is increased by real-time cross-correlation of the raw data acquired by each receiver with a pilot trace of the emitted pulse trains. The cross-correlation process also converts the long pulse trains emitted by the source into the short (broadband) pulses needed for resolution of multiple seismic waves arriving within a short period of time.

At Mactaquac, each 8-channel PS8R record was generated from a cross-correlated stack (average) of 490 pulse trains—the maximum number that could be averaged. Each record, acquired at a sample rate of 0.002 ms and 5–8 ms in length, took about 1 min to acquire. Up to five records were acquired at each depth to allow for further signal averaging and noise suppression. The tool was then unclamped and raised to the next desired depth.

Figure 8 shows a photograph of the 4 m long PS8R tool as deployed at Mactaquac. The source module was situated at the bottom. Eight receivers spaced 15 cm apart were located 1.185–2.235 m above the source. The source and each receiver were

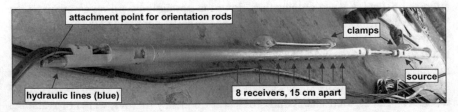

Fig. 8 Photograph of the clamping, 8-channel Vibrometric PS8R borehole seismic system assembled on surface at Mactaquac Dam

coupled to the borehole wall through hard rubber pads aligned vertically on one side of the tool, opposite to clamps which are hydraulically extended against the other side of the hole with a pressure of approximately 20 bars. The source transducer is designed to vibrate in a direction perpendicular to the face of its pad. Similarly, the receivers are designed to be sensitive predominantly to incoming vibrations perpendicular to their pads.

The PS8R tool was operated in two different orientations at Mactaquac in order to preferentially emit either P-waves or S-waves in the direction of the concrete-clay interface (Enescu and Cosma 2010). We sought to emphasize P-wave reflections by operating the instrument with its source and receiver pads pressed against the high side of the borehole—i.e. pointing towards the concrete/clay interface or wall. Conversely, S-wave reflections were emphasized by rotating the tool 90° so that the pads were pointing towards the headpond—i.e. in a direction parallel to the concrete/clay interface. These two modes were called "wall-perpendicular" and "wall-parallel" respectively. The orientation of the tool was controlled manually by the use of 2 m long aluminum orientation rods (Fig. 9), which interlocked with each other and with the top of the tool. The rods, marked with depth labels, were also used to manually lower and raise the tool in the borehole. The tool was secured at a given depth by use of a vise (Figs. 9) which closed around the circumference of the uppermost rod. Hydraulic pressure was then applied to clamp the tool to the sides of the borehole.

The original survey plan called for wall-perpendicular and wall-parallel measurements to be made in two separate runs at depth increments of 30 cm. However, in order to ensure that the tool stayed above the water, which was rising at a rate of approximately 7 cm/min, it was decided to survey the hole four times at 60 cm spacing (once at even stations and once at interleaved odd stations for each orientation).

Fig. 9 Operation of the PS8R borehole seismic system in borehole SEPI-1 at Mactaquac, August, 2012

Ultimately, we had time for three production runs. A submersible pump was used to remove water from the hole between survey runs.

Two unexpected challenges were encountered during data acquisition and subsequent processing of the PS-8R reflection data. First, despite being acquired in dry boreholes, there was significant interference from relatively slow direct waves—presumably surface waves propagating along the walls of the dry borehole—which travelled up the borehole wall and overprinted any interface reflections. This is evident in Fig. 10 which shows, on a typical shot record, where S-wave reflections from the concrete/clay interface would be expected to appear.

Second, it became apparent that signal strengths were critically dependent on how firmly the source and receiver pads were coupled to the borehole wall; signals acquired during the initial wall-parallel production run and the following wall-perpendicular production run were very weak, suggesting that the pistons on the hydraulic clamps did not have sufficient range of motion (nominally 5 mm) to press

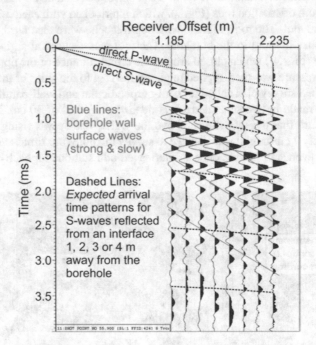

Fig. 10 Typical PS8R shot record acquired in the wall-parallel orientation with source at 9.2 m depth, following application of a 1–28 kHz causal bandpass filter. Arrival of the direct S-wave, with a velocity of 2415 m/s, is clearly visible across all 8 receivers. Arrival times that would be expected for S-wave reflections from distances of 1, 2, 3, and 4 m are shown as dashed black lines. Unfortunately, any S-wave reflections from the SEP interface are obscured by stronger arrivals, interpreted to be borehole wall surface waves or borehole guided waves that sweep across the record more slowly than the shear wave, despite the fact that the PS8R is operating clamped to the sides of an air-filled borehole. There is only weak evidence of a direct P-wave arrival at the start of the record; this is to be expected because the receivers are insensitive to particle motion parallel to the borehole wall

firmly against the concrete. Thin washers (up to 1.5 mm thick) were inserted as spacers on the ends of the pistons in order to increase their extension. This improved a subsequent production run in wall-parallel mode, but the wall-perpendicular signal strength remained weak, leading to speculation that the hole was not exactly circular in cross-section, perhaps due to the fact that it was inclined and had been grouted and re-drilled multiple times.

Data Processing

Although reflections from the concrete/clay interface were not obvious in the shot records, it was expected, given the very small dip of the interface relative to the borehole (Fig. 4), that they would be enhanced by applying the relatively simple multi-channel common midpoint (CMP) stacking technique used successfully for surface seismic reflection surveying over gently dipping layers. More sophisticated reflection imaging techniques, including pre-stack migration had been used in the oil and gas well case studies cited above, and those methods would offer advantages, especially for imaging interfaces at a high angle to the borehole. However, those studies also benefited from greater data volumes (more receivers, finer shot spacing) and longer source-receiver offsets. It is not clear whether the more sophisticated imaging methods would be justified in this case though it remains a question worth exploring.

Originally, we had hoped to acquire PS-8R data for both P-wave and S-wave reflection imaging of the concrete/clay interface using shotpoints at 30 cm increments in the borehole. However, as mentioned above, two of the three production runs—one for P-waves and one for S-waves—appeared to suffer from poor coupling between the tool and the borehole wall. Results obtained by processing those data were non-conclusive and are not presented. We focus instead on data acquired during the third production run, conducted in wall-parallel mode favouring the reception of S-wave reflections from the concrete/clay interface.

Velocity Models

P-wave and S-wave velocities as a function of depth in the concrete SEP were determined from the FWS logging data in the standard way by semi-automated semblance analysis on the direct P and S-wave arrivals recorded by four receivers spaced 1 foot apart, 3–6 feet above the source. The raw Vp and Vs logs are shown in Fig. 11, along with smoothed values calculated using symmetric moving averaging windows of various lengths. While some of the fine-scale variability is likely noise, variations with wavelengths on the order of a metre and more are quite likely associated with changes in concrete properties. Such variability is to be expected given that the SEP was constructed of near-horizontal lifts that were nominally 5 ft or 10 ft thick (see

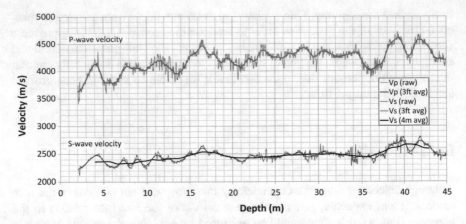

Fig. 11 P-wave and S-wave velocities measured in borehole SEPI-1 by full waveform sonic (FWS) logging. Depth refers to slant depth, measured along the inclined borehole

Fig. 4). Vp and Vs both exhibit a gradual increase with depth and their shorter wavelength variations are quite highly correlated, although there are sufficiently large variations in the Vp/Vs ratio to cause Poisson's ratio to range from approximately 0.20 to 0.27. Average P and S-wave velocities are approximately 4210 and 2470 m/s with nearly all measurements falling within 10% of those averages.

The observed range of FWS velocities in Fig. 11 was comparable to those measured at much lower frequencies (100–500 Hz) in surface-to-borehole (vertical seismic profile) measurements made using a sledgehammer source on the concrete deck; P-wave velocities measured using a 12-channel hydrophone eel in the borehole were just slightly lower at 3900–4200 m/s, while the average S-wave velocity measured using a downhole 3-component geophone was 2500 m/s, in excellent agreement with the average from FWS logging.

The P and S-wave velocity models were assumed to be 1-dimensional, varying only with distance along the borehole, meaning that they were assumed to be composed of layers perpendicular to the borehole. Ideally, the model would have been adjusted to be 2-dimensional, with layers oriented parallel to the concrete lifts that lie at an angle of approximately 30° to the inclined borehole. However, the effort that would have been required to incorporate such a 2D velocity model was not considered to be worthwhile in light of the rather small velocity contrasts between layers and the minimal improvement to seismic imaging that would be expected.

Seismic Reflection Survey Processing

Pre-stack Processing

The PS-8R data were processed using VISTA (v. 12) seismic reflection processing software. Key steps applied during the initial stages of data processing, leading to production of CMP stack seismic section, included bandpass filtering, trace amplitude normalization, normal moveout (NMO) correction, and CMP stacking. Details of the pre-stack processing sequence are given in Table 1.

Bandpass filtering from 7 to 16 kHz (Fig. 12) was the sole pre-stack process used to combat borehole surface wave noise. Attempts to further reduce surface wave noise in the shot records through velocity-based multi-channel filtering procedures such as f-k (frequency-wavenumber) filtering were not effective, because the low velocity of

Table 1 Pre-stack processing flow

Step	Description
(i) Data import	69 shot records in SEGY format, acquired in wall-parallel mode at 60 cm intervals spanning shot depths of 44.035–3.235 m, were imported into the processing software. The temporal sampling interval (2 μs) and all shot/receiver positions within headers were multiplied by 10 to prevent apparent loss of numerical precision within some processing commands likely caused by round off error, given that the software was designed for much larger scale, lower frequency seismic surveys
(ii) Trace editing	Removed small DC bias from traces, muted time-zero noise, and edited shot records to remove a small number of dead or excessively noisy traces
(iii) Bandpass filtering	A 7–16 kHz minimum phase Butterworth bandpass filter (applied as 0.7–1.6 kHz to the scaled data) was used to preserve most of the expected bandwidth of reflected S-wave arrivals and attenuate low frequencies dominated by surface waves travelling along the borehole wall (Fig. 12)
(iv) Trace normalization	Trace amplitudes were normalized by their mean absolute values over the window 0.4–2.5 ms to balance amplitudes from one trace to the next and one shot location to the next
(v) Top mute	A top mute was applied to remove direct P- and S-wave arrivals in each shot record
(vi) NMO correction	Normal moveout (NMO) corrections were applied to each shot record, using the 4 m running average FWS S-wave velocity model, in order to flatten any S-wave reflections
(vii) CMP stack	Stack (average) channels 1–7 in each NMO-corrected shot gather to enhance the strength of reflected arrivals relative to other seismic events. (Omitting channel 8—the receiver closest to the source—resulted in a slightly improved stack)
(viii) CMP section	Plot the 69 stacked traces (one for each shot record) side by side to generate a seismic section

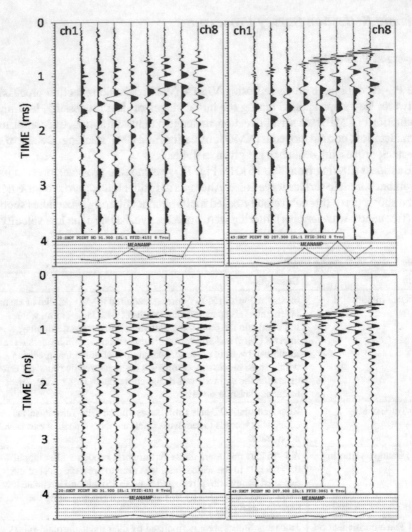

Fig. 12 *Upper row*: Two sample PS8R shot records at depths 14.6 m (left) and 32.0 m (right) after processing to remove small DC bias and mute time-zero noise. *Lower row*: Same two shot records after 7–16 kHz minimum phase Butterworth bandpass filtering with gentle roll-off slopes of 12 dB/octave. The amplitude of each trace has been normalized by its mean for display. Mean amplitudes of each trace are graphed at the base of each record, using a common scale

the surface wave arrivals caused them to appear spatially aliased despite fact that the receivers were only 15 cm apart. We also noted that the bandpass-filtered shot records, such as those shown in Fig. 12, exhibited a reverberatory character, in addition to waveform variability presumably caused by changes in source and receiver coupling. Deconvolution or other spectral broadening processes might be worth investigating in future work to improve waveform definition and stability.

The pre-stack processing sequence described in Table 1 is essentially the same as that used to produce a common midpoint (CMP) stack seismic section when working with multichannel seismic reflection data acquired on surface. However, because the number of channels/receivers was relatively low, and because shot spacing was rather large compared to the receiver spacing, the CMP bins were made 60 cm wide (equal to the shot spacing) so that each bin would include traces from all 8 source-receiver offsets. This made each CMP gather equivalent to a shot record. The reflection point smearing associated with this approach was considered acceptable given that the dip of the interface relative the borehole was expected to be only about 3.7° (see Fig. 4).

Post-stack Processing

The CMP stack seismic section (Fig. 13) generated in this way was dominated by the borehole surface waves evident in the shot records. Surface waves travelling along the borehole wall directly from the source to the receivers appear as strong horizontal banding in the seismic section because their arrival times were largely independent of position in the borehole. Although they did not sum constructively across traces during CMP stacking, they remained stronger than any reflection from the concrete/clay interface which would be expected to appear as a dipping feature, closest to the borehole at surface and most distant at depth. There are hints of a dipping feature following the expected trend for an S-wave reflection (dashed line in Fig. 13) but there is clearly a need to attenuate the direct surface wave noise more than was achieved by frequency filtering and CMP stacking alone. To this end, a 7-trace (4.2 m long) horizontal running average operator was applied to the CMP stack to generate an estimate of the direct surface wave noise (Fig. 14). This noise estimate was then subtracted from the CMP stack yielding the dramatically different and more informative "mean-filtered" CMP section shown in Fig. 15.

Next, the mean-filtered seismic section was subjected to 2D migration (Fig. 16) using the 4 m running average S-wave velocity model, in an effort to improve S-wave reflection continuity by collapsing diffraction patterns from small discontinuities and to correct (steepen) the dip of reflection events. The process assumed that the velocity model was correct and that there was no significant scattering from out-of-plane (3D) features. Both assumptions could be questioned, particularly since the scattered wavefield likely includes P-waves in addition to the S-wave interface reflection being sought. The effects of migration in this case were subtle, although it did suppress some clutter which may have been a consequence of diffractions.

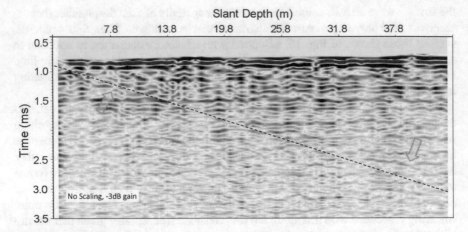

Fig. 13 CMP stack seismic section generated using channels 1–7 after 7–16 kHz bandpass filtering. Horizontal banding, representing direct waves having arrival times nearly independent of depth, dominates the section. Dashed line shows approximate arrival time that would be expected for an S-wave reflection from the concrete/clay interface. Arrows highlight semi-coherent seismic events that may be the reflection

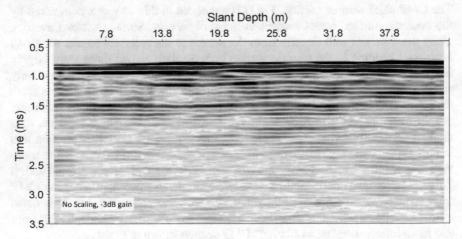

Fig. 14 Seven-trace (4.2 m) running average mean of the CMP stack seismic section. This was taken as an estimate of the direct surface wave noise and subtracted from the CMP stack to yield the mean-filtered CMP stack section in Fig. 15

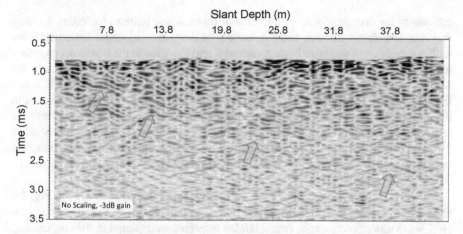

Fig. 15 Seven-trace mean filtered CMP stack section (i.e. section in Fig. 13 minus section in Fig. 14). Arrows highlight coherent segments of a seismic event that may be an S-wave reflection from the concrete/clay interface

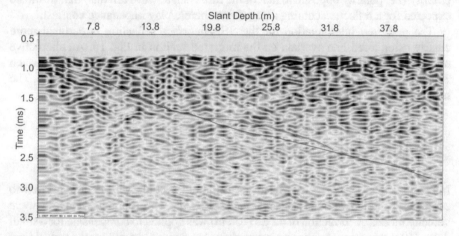

Fig. 16 Seven-trace mean filtered CMP stack section after the application of 2D migration. Red and green lines show preferred and alternative interpretations for the S-wave reflection from the concrete/clay interface

Interpretation and Recommendations

Seismic Section Interpretation

The mean-filtered CMP-stack seismic section in Fig. 15, acquired, and processed in such a way as to favour imaging of a S-wave reflection from the concrete/clay interface, shows several coherent to semi-coherent seismic arrivals, some of which exhibit continuity for distances of several metres or more. Their pattern, including the way that they interfere with each other indicate that there are numerous heterogeneities within the concrete that have back-scattered S and P-waves to the borehole, probably from multiple directions. These heterogeneities could include the boundaries of the SEP structure as well as the near-horizontal contacts between concrete pour lifts and fractures within the concrete. Given the strength of direct borehole surface waves in the shot records, it is also highly likely that some of the coherent seismic arrivals represent back-scattering of surface waves from concrete pour lifts, fractures or other discontinuities intersected by the borehole. The dominant frequency of the coherent events is approximately 7 kHz, representing a wavelength of 60 cm for P-waves or 35 cm for S-waves.

Notably, there is one prominent seismic arrival, highlighted by arrows in Fig. 15, that exhibits arrival times and dip very comparable to those that would be expected for an S-wave reflection from the concrete/clay interface as depicted in the engineering plan (Fig. 4). The reflection is nearly continuous to a depth of 18 m in the borehole, and can be extrapolated through two more semi-continuous segments to the bottom of the survey at 43 m depth. It is nearly linear, as expected, and it has the inverted polarity (i.e. polarity opposite to that of the direct shear wave arrivals) that would be expected for a reflection coming a negative concrete/clay impedance contrast.

The preferred interpretation for the SEP interface reflection is outlined more clearly using a red line overlain on the migrated section in Fig. 16. An alternative interpretation, highlighted by a green line, differs only in the depth range between 20 and 32 m where a slightly earlier and more coherent event has been selected. The red line interpretation is preferred because it is more linear.

Quantitative Estimate of Interface Position

Interface reflection time picks from the migrated section were next converted to estimates of borehole-interface separation using the 4 m running average S-wave velocity model. A correction of 42 μs (~5 cm) was applied to compensate for a small delay (10 μs) imposed by the minimum phase bandpass filter and for the arrival time difference (~32 μs) between the onset of the SEP reflection and the waveform peak that was actually picked. The calculated separations ranged from approximately 0.9 to 3.6 m. Finally, the estimated interface position was plotted as shown in Fig. 17, taking into account a survey of the borehole inclination, which was slightly steeper (0.2° steeper on average) than the 60° that had been intended.

Figure 17 has been rotated so that the green "datum line" is inclined at an angle of 60°—the intended inclination for borehole SEPI-1. The slightly steeper borehole, which deviates most below 20 m depth is shown in orange. The solid blue line shows the position that would be expected for the concrete—embankment interface according to the engineering design plan (Fig. 4), on the understanding that the borehole was collared 1 m away from interface at surface. The solid black dots and the open brown dots (which overlay each other over most of the length of the hole)

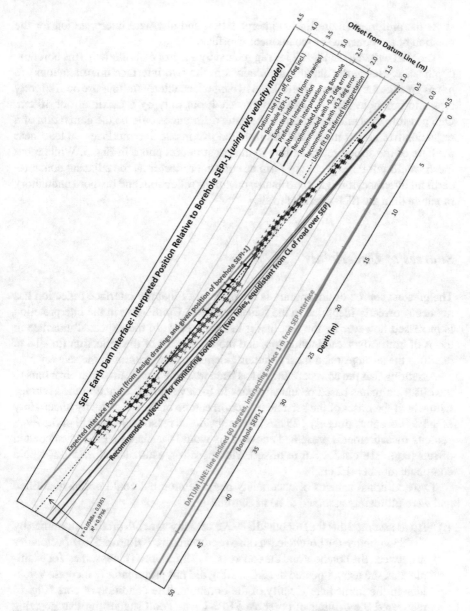

Fig. 17 Interpreted position of the concrete—embankment dam interface (black line) relative to the surveying borehole SEPI-1 shown in orange at its true inclination. Recommended trajectory for the two seepage monitoring boreholes is shown in red, with grey lines illustrating how it would vary for ±0.2° inclination error. Note that the offset scale is exaggerated by a factor of four relative to the borehole slant depth scale

show the preferred (more linear) interpretation and alternate interpretation for the position of the concrete—embankment interface.

The "Offset" scale in Fig. 17 is exaggerated by a factor of 4 relative to the borehole depth scale. Hence the angular undulations in the two interface interpretations are not as great as they appear. Overall, the two interface interpretations are both slightly closer to borehole SEPI-1 than was expected, but only by a maximum of 40 cm. One possible explanation for this systematic difference could be the generation of a reflection from rebar mesh that was installed just inside the interface—at least near the base of the structure, according to the construction photo in Fig. 5. While rebar mesh would not be expected to form as strong a reflector as the adjacent concrete-earth fill interface, it might contribute a precursor reflection, and be more undulatory in nature than the SEP interface itself.

Sources of Uncertainty

The greatest source of uncertainty is the possibility that the interface reflection has not been correctly identified in the seismic section. Confidence in the interpretation is increased however by the fact that it lies very close to the expected interface in terms of both offset and inclination, and the upper part of the reflection (to ~18 m depth) is the strongest and most coherent seismic arrival present in the section.

Assuming that the correct reflector has been identified, we can assign error bars to the offset estimates based on uncertainties in S-wave velocities and in the accuracy with which the onset of the seismic reflection can be picked. A velocity uncertainty of ±5% (i.e. approximately 125 m/s) was considered reasonable in light of the FWS velocity measurements presented above. This would translate to ±5% in calculated offsets (e.g. ±18 cm for 3.6 m offset). Picking error, estimated at ±15 μs, would contribute another ±2 cm.

Two additional sources of uncertainty bear mention but could not be quantified and were ultimately assumed to be negligible:

(i) It was assumed that the smoothed S-wave velocity vs depth profile determined by FWS logging in the borehole was also representative (within ±5%) of velocities in between the borehole and the concrete/clay interface. This assumes, for example, that the grout injected during drilling did not preferentially increase velocities in the immediate vicinity of the borehole. The condition of core (Fig. 6) recovered the drilling of borehole SEPI-1 was good suggesting that grouting was not needed to fill extensive fracturing, although it may have reduced inter-granular porosity immediately around the hole. The fact that a long wavelength surface to borehole S-wave velocity measurement was in excellent agreement with the FWS logging estimates provides some evidence that borehole logging measurements were representative of those throughout the concrete mass but cannot rule out the possibility of off-hole low velocity zones.

(ii) An error in the intended easterly azimuth of the hole as it was drilled could have caused the interface to be closer than expected at depth if it was large enough to cause the hole to drift upriver or downriver by distances exceeding 1.1 or 4.4 m respectively, beyond which there are slight changes in the orientation of the SEP wall. However, this was considered unlikely given the care with which the hole was targeted and drilled and the large diameter (and hence stiffness) of the drill rods. By way of comparison, the average 0.2° error in intended inclination resulted in the bottom of the borehole being only 18 cm off of the original target.

Installation of Seepage Monitoring Boreholes

The seepage monitoring plan for the concrete/clay interface called for installation of two boreholes to be drilled from surface parallel to the interface and as close to it as possible to maximize the sensitivity of temperature and SP monitoring sensors to any interfacial flow. Instrumentation to be installed in the holes included a heat trace cable, two fibre optic DTS systems (one in each hole), piezometers, and an array of electrodes for SP monitoring (Butler et al. 2014). The two holes, separated by 3.3 m were to be drilled parallel to one another on either side of the centerline of the dam-crest road and equidistant from the interface so that the thermal signals generated by active heating in one borehole would be carried downgradient to the second borehole by any concentrated seepage along the interface.

The benefit of proximity to the concrete/clay contact had to be balanced against the importance of avoiding damage by inadvertently drilling across it. Helpfully, the sensitivity of borehole temperature monitoring during active heating, to seepage along a nearby interface, was the subject of numerical modelling carried out in a companion project (Shija and MacQuarrie 2015). One of the key results was that temperature anomalies expected opposite a zone of concentrated seepage, in a borehole undergoing active heating, would be up to five times greater at a borehole—interface offset of 0.5 m compared with an offset of 1.0 m. Considering that result, and the error bars and undulations in the estimated interface position, it was therefore recommended that the two parallel monitoring boreholes be installed within 0.5 m of the interface.

The red line in Fig. 17 shows the linear trajectory recommended for the two monitoring holes. It was set to be 95% of the distance from the green datum line to the dashed black line defining the best linear fit to the "preferred interpretation", less an additional cushion of 35 cm. This ensured that the trajectory did not cross either of the two interface interpretations or their error bars. Its closest approach to an error bar was 0.11 m. The distance from the recommended monitoring borehole trajectory to the best linear estimate for the SEP interface (dashed black line) was within the desired range, increasing from 0.30 m at surface to 0.50 m at the foot of the interface (48 m depth).

During the late summer and fall of 2013, the recommended monitoring boreholes were drilled and continuously cored with the same diamond drill and heavy rods as previously used for the installation of the survey hole SEPI-1 (Fig. 16). The drill core was entirely concrete. Two pieces of rebar tie rods were observed in core samples—likely representing pieces of the tie rods seen extending inward from the SEP interface in the construction photo, Fig. 5. The holes were fitted with instrumentation in early November, 2013, and (together with other sensors on the dam surface) have provided a wealth of temperature (DTS) and SP data through all seasons (e.g. Ringeri et al. 2016; Yun et al. 2016). This has been of use to NB Power in seepage surveillance and is contributing to ongoing research into seepage monitoring technologies. Amongst other results, modelling has shown that at least some of the temperature anomalies observed in the DTS data are likely caused by seepage along fractures in the concrete rather than seepage along the concrete—embankment interface (Yun et al. 2018); this has been confirmed by "erasure" of such anomalies following targeted remedial grouting within the concrete (Yun 2018).

Discussion and Conclusions

As dams age and require more inspection or remediation, there will be increased call for methods that can be used to detect interior defects, or confirm their internal structure for purposes of installing piezometers, DTS cables and other types of monitoring instrumentation. The objective of this study was to confirm the location of the steeply inclined interface between an embankment dam and a concrete diversion sluiceway as accurately as possible for purposes of installing seepage monitoring instrumentation. Given a prohibition against drilling into the dam's clay till core, GPR or high frequency seismic methods deployed from a survey borehole drilled into the concrete sub-parallel to the interface 1–4 m away were selected as the methods most likely to yield adequate precision. After GPR failed to achieve sufficient range of penetration, the focus switched to seismic methods.

To our knowledge, this is the first report of single borehole seismic reflection imaging applied in a dam or any concrete structure. A novel 8-receiver wall-locking seismic tool, was used in two different orientations to collect one data set favouring the reception of P-wave reflections and another favouring reception of S-wave reflections from the concrete/earth dam interface, though the former suffered from inadequate tool clamping. Despite operating in a dry borehole, surface waves travelling directly up the borehole dominated the 8-channel shot records during the times when S-wave reflections were expected. Given the shallow dip (~3.7°) expected between the borehole and the target interface, a relatively simple processing flow consisting of bandpass filtering, CMP stacking and mean trace subtraction was used to attenuate the direct surface waves and reveal an interpreted S-wave reflection exhibiting a dip and offset very comparable to that expected for the concrete—embankment interface.

Interpretation of the S-wave reflection section was complicated by the presence of residual borehole surface wave noise, and by multiple semi-coherent arrivals dipping both away from and towards the borehole. These events may represent scattering of S-waves or P-waves from off-hole heterogeneities, or of borehole surface waves from heterogeneities intersecting the borehole (such as the concrete pore lift boundaries and fractures). It would have been helpful to have shots more closely spaced than 60 cm to reduce the chances of spatial aliasing of such back-scattered signals, and to allow one to take advantage of more sophisticated pre-stack migration imaging techniques that could have reduced some of this clutter. The strength and coherence of the concrete—embankment interface reflection might also be improved by processing efforts to reduce waveform variability attributed to differences in source or receiver coupling.

With respect to the instrumentation, the PS-8R borehole tool would have benefited from clamps with a larger range of motion. As shown in Fig. 10, an ability to shift the receivers to longer offsets would have been beneficial to reduce the overlap between interface reflections and guided waves for the 1–4 m range of interface distances in this case. Also, despite the fact that receivers were only 15 cm apart, high frequency and slow velocity of surface wave arrivals resulted in spatial aliasing which prevented them from being removed by f-k (frequency—wavenumber) or other types of velocity filtering. Tighter receiver spacing would be necessary to overcome that problem. As an alternative to adding receivers, instrument designers could add one or more additional sources to the tool, each of which would provide a different set of source-receiver offsets that could be combined during data processing to yield virtual/composite shot records offering closer receiver spacing and a wider range of offsets.

Despite the interference evident in the final stacked and migrated S-wave sections, prior information on the dam's expected internal structure helped instil confidence in the quantitative interpretation developed for the concrete/clay interface position. Two 50-m long boreholes for seepage monitoring instrumentation have since been installed within an estimated 50 cm of the contact. Given the potential for improvements in data acquisition and processing, prospects for further development of seismic imaging in dams and other concrete infrastructure appear to be promising.

Acknowledgements UNB students Andrew Ringeri and Jessie Brown provided assistance in the field and in processing the FWS data. Vibrometric Field Engineer Cristian Vasile provided expertise in acquisition of the PS-8R data. We thank John Fletcher and Ian Campbell of NB Power for logistical support. This project was part of research into seepage surveillance technologies funded jointly by NB Power and the Natural Sciences and Engineering Research Council (NSERC) through a Collaborative Research and Development Grant. The manuscript was improved through the helpful recommendations of two anonymous reviewers.

References

Bing, W., Guo, T., Hua, W., and Bolei, T., 2011, Extracting near-borehole P and S reflections from array sonic logging data, *Journal of Geophysics and Engineering*, **8**, 308–315.

Butler, K.E., Ringeri, A., MacQuarrie, K.T.B., Shija, N.P., Colpitts, B.G., and McLean, D.B., 2014, Installation and early performance of a seepage surveillance system at the Mactaquac Dam, New Brunswick, Symposium on Application of Geophysics to Eng. and Env. Problems (SAGEEP 2014), Mar. 16–20, Boston.

American Society of Civil Engineers, 2017, 2017 Infrastructure Report Card – Dams, https://www.infrastructurereportcard.org/cat-item/dams/ accessed July 29, 2018.

CEATI International Inc., 2018, Dam Safety Interest Group, https://www.ceati.com/collaborative-programs/generation/dsig-dam-safety/ accessed July 29, 2018.

Chabot, L., Henley, D. C., Brown, R. J., and Bancroft, J. C. 2002, Single-well seismic imaging using full waveform sonic data: An update. SEG Technical Program Expanded Abstracts: pp. 368–371.

Conlon, R. and Ganong, G. 1966, The foundation of the Mactaquac rockfill dam, *Engineering Journal*, April: 33–38.

Cosma, C. 1995, Ultra Acoustic Tomography for Characterizing the Excavation Disturbed Zone, 57th EAGE Conference and Technical Exhibition, Glasgow, Scotland.

Cosma, C., Enescu, N., Powell, B. and Wood, G., 2006, Structural mapping for uranium exploration by borehole seismic, EAGE Near Surface 2006 conference, Helsinki, Finland.

Cosma, C., Enescu, N. and Heikkinen, E., 2010, Very high resolution hard rock seismic imaging for excavation damage zone characterization, EAGE Near Surface 2010 conference, Zurich, Switzerland.

Cosma, C., Wolmarans, A., Eichenberg, D. and Enescu, N., 2007, Kimberlite delineation by seismic side-scans from boreholes. Workshop 6 paper, Exploration '07, Toronto, Canada, Publicly accessible at http://www.dmec.ca/ex07-dvd/E07/proceedings.html.

Daley, T.M., Gritto, R., Majer, E.L., and West, P., 2003, Tube-wave suppression in single-well seismic acquisition, *Geophysics*, **68**, 863–869.

Emersoy, C., Chang, C., Tichelaar, B., and Quint, E., 1998, Acoustic imaging of reservoir structure from a horizontal well, *The Leading Edge*, **17**, 940–946.

Emsley, S., Olsson, O., Stenberg, L., Alheid, H.J. and Falls, S., 1997. ZEDEX – A study of damage and disturbance from tunnel excavation by blasting and tunnel boring, SKB Technical Report 97–30, Publicly accessible at http://www.skb.com/publications/.

Enescu, N. and Cosma, C., 2010, EDZ Seismic Investigations in ONKALO, 2009. Factual Report Posiva Workreport 2010-26, 114 pages, Publicly accessible at http://www.posiva.fi/en/databank/workreports.

Federal Emergency Management Agency, 2005, The National Dam Safety Program Research Needs Workshop: Seepage through Embankment Dams, FEMA report 535, https://www.fema.gov/media-library/assets/documents/1033?id=1452 accessed July 29, 2018.

Federal Emergency Management Agency, 2015, FEMA National Dam Safety Program Fact Sheet, FEMA P-1069, https://www.fema.gov/media-library/assets/documents/5865 accessed July 29, 2018.

Federal Energy Regulatory Commission, 2017, Engineering Guidelines for the Evaluation of Hydropower Projects, Chapter 14: Dam Safety Performance Monitoring Program, Appendix J: Dam Safety Surveillance and Monitoring Plan Outline, https://www.ferc.gov/industries/hydropower/safety/guidelines/eng-guide/chap14.pdf, accessed July 29, 2018.

Fell, R., MacGregor, P., Stapledon, D., and Bell, G. 2005, Geotechnical Engineering of Dams, A.A. Bakema Publishers, Leiden, The Netherlands.

Franco, J.L.A., Ortiz, M.A.M., De, G.S., Renlie, L., and Williams, S., 2006, Sonic investigations in and around the borehole, Oilfield Review, **18**, 14–33.

Gilks, P., May, T. and Curtis, D. 2001, A review and management of AAR at Mactaquac Generating Station, Proceedings of the Canadian Dam Association 2001 Annual Conference, pp. 167–177.

Giroux, B., Butler, K.E., and McLean, D.B., 2011, Borehole radar investigation at the Mactaquac Dam, Proceedings of the Canadian Dam Association 2011 Annual Conference, Oct. 15–20, Fredericton, NB, 10 pp.

Greenwood, A., J. C. Dupuis, H. Abdulal, and Kepic, A. 2012, Rigid corrugated baffle system for tube-wave suppression in deep boreholes, 82nd Annual International Meeting, SEG, Expanded Abstracts, https://doi.org/10.1190/segam2012-0972.1.

Hornby, B.E., 1989, Imaging of near-borehole structure using full-waveform sonic data, *Geophysics*, **54**, 747–757.

Mattsson, G., Hellstrom, J.G.I. and Lundstrom, T.S. 2008, On Internal Erosion in Embankment dams, Research report, Vol. 14, 1402–1528, Lulea University of Technology, Sweden.

Milligan, P.A., Rector, J.W., and Bainer, R.W., 1997, Hydrophone imaging at a shallow site, *Geophysics*, **62**, 842–852.

Ringeri, A., Butler, K.E., and McLean, D.B., 2016, Long term monitoring and numerical modeling of self-potential for seepage surveillance at Mactaquac dam, New Brunswick, Canada. Paper 3901, Proceedings of GeoVancouver 2016, the 69th Canadian Geotechnical Conference, Oct. 2–5, Vancouver, 8 pp.

Shija, N.P. and MacQuarrie, K.T.B., 2015, Numerical simulation of active heat injection and anomalous seepage near an earth dam–concrete interface, *Int. J. Geomech.*, **15**, https://doi.org/10.1061/(asce)gm.1943-5622.0000432.

Tang, X.M., Zheng, Y., and Patterson, D., 2007, Processing array acoustic-logging data to image near-borehole geologic structures, *Geophysics*, **72**, E87-E97.

Tawil, H., and Harriman B. 2001, Aquifer Performance Under the Mactaquac Dam, Proceedings of the Canadian Dam Association 2001 Annual Conference, pp. 99–109.

Yun, T., 2018, Investigation of Seepage near the Interface between an Embankment Dam and Concrete Structure: Monitoring and Modelling of Seasonal Temperature Trends, MSc Thesis, University of New Brunswick.

Yun T., Butler, K.E., MacQuarrie K.T., Mclean, B., & Campbell, I., 2018, Seasonal temperature monitoring and modelling for Seepage Reconnaissance in an Embankment Dam. Extended abstract, 24th European Meeting of Environmental and Engineering Geophysics, EAGE, 5 pp., Sep. 9–13, 2018, Porto, Portugal.

Yun, T., Ringeri, A., Butler, K.E., and MacQuarrie, K.T.M., 2016, Seepage Reconnaissance in an Embankment Dam using Distributed Temperature Sensing (DTS): Monitoring and Modelling of Seasonal Effects, Poster NS33B-1978, AGU Fall Meeting, Dec 12–16, San Francisco.

Self-potential Imaging of Seepage in an Embankment Dam

A. Bouchedda, M. Chouteau, A. Coté, S. Kaveh and P. Rivard

Abstract We have investigated seepage in Les Cèdres embankment dam (Valley-field, Canada) by combining self-potential tomography (SPT), electrical resistance tomography (ERT), thermometry, electromagnetic (EM) conductivity and magnetic measurements. SPT consists of inverting self-potential data to retrieve the source current density distribution associated with water flow pathways (streaming current density) in embankment dams. The SPT inverse problem relies on the resistivity model of the dam that is obtained by 3-D ERT. Our 3-D SPT code is based on Occam's inversion. The forward problem is solved using the finite-volume scheme. The investigated embankment dam is used to channel water from the Saint-Lawrence River to a hydroelectric plant. It separates in its upstream and downstream sides Les Cèdres and St-Timothée reservoirs respectively. St-Timothée reservoir is emptied during the winter and filled during the summer. Temperature monitoring was done in a borehole installed in the middle of the survey zone. To build a better understanding of water flow through the dam, it is important to separate the part of the source current density caused by the electrokinetic effect from the other sources (principally electro-chemical). In order to achieve that, ERT, EM31, magnetic and thermometric measurements have been used in the interpretation. EM conductivity maps allowed identifying two linear anomalies caused by metal-shielded electrical cables. The mag-

A. Bouchedda
Centre Eau Terre Environnement,
Institut national de la recherche scientifique, Quebec City, QC, Canada
e-mail: Abderrezak.bouchedda@ete.inrs.ca

M. Chouteau (✉)
Ecole Polytechnique de Montreal, Montreal, QC, Canada
e-mail: michel.chouteau@polymtl.ca

A. Coté · S. Kaveh
Institut de recherche d'Hydro-Québec, Varennes, QC, Canada
e-mail: alain.cote@ireq.ca

S. Kaveh
e-mail: saleh.kaveh@ireq.ca

P. Rivard
Université de Sherbrooke, Sherbrooke, QC, Canada
e-mail: patrice.rivard@USherbrooke.ca

© Springer Nature Switzerland AG 2019
J. Lorenzo and W. Doll (eds.), *Levees and Dams*,
https://doi.org/10.1007/978-3-030-27367-5_4

69

netic survey shows an important anomaly zone that is probably related to a metallic object. Therefore, all measurements near these zones were discarded from inversion. SPT shows a few seepage sources on the upstream dam side at a depth between 4 and 5 m. Two of them are confirmed by geotechnical testing. The water flow through the dam appears complex. It is partly controlled by a permeable zone that is well identified in the resistivity model. In the vicinity of the vertical temperature profile the SPT shows that the water flows parallel to the dam orientation and not through the borehole used for thermometry. This is why there is no clear indication of water seepage in temperature measurements. Finally, all observable seepage outlets on the downstream side can be related to the SPT anomalies and are observed as conductive zones in the resistivity model.

Introduction

Embankment dams experiencing seepage or leakage are exposed to the development of internal erosion which represents a high risk of dam failure. Internal erosion is caused by the water flow with particle transport. When seepage are observed at the ground surface of the dam, geotechnical and geophysical investigations are typically performed (e.g. Butler et al. 1990). A few geophysical techniques have been developed to detect preferential fluid flow pathways in embankments (Ogilvy et al. 1969; Haines 1978; Gex 1980; Al-Saigh et al. 1994; Panthulu et al. 2001; Rozycki et al. 2006; Rozycki 2009; Sheffer 2002, 2007; Sheffer and Oldenburg 2007; Bolève et al. 2009, 2012; Bolève 2013). Both passive and active electric and thermometric methods have proven to be effective and are widely used (Sjödahl et al. 2008; Smith et al. 2009). In the last decade, the spontaneous potential tomography (SPT) was probably the technique that experienced the most active development because the SP signal is directly sensitive to water flow through a porous media (Sheffer and Oldenburg 2007; Bolève et al. 2007; Jardani et al. 2008; Revil and Jardani 2013). For example, Bolève et al (2007), Sheffer and Oldenburg (2007) studied the sensitivity of the SP response to the flow and dam properties and geometry for the cases of a saturated and an unsaturated medium.

Les Cèdres is an embankment dam that guides water from a channel of the St-Lawrence River to a power plant. It was built in 1914 and it has been plagued with seepage for many years. Water leakage can be observed at dam toe at a few places. In addition, a few water infiltrations zones were highlighted using geotechnical investigations (Qualitas 2010) and magnetometric resistivity (MMR) technique (Kofoed et al. 2011). For these reasons, Les Cèdres was chosen by Hydro-Quebec, the dam owner and operator, to evaluate non-destructive methods that could be applied to other problematic dams to detect seepage and assess remediation work. The main objective was to give a good understanding of water flows within the dam to plan an effective grouting program in order to stop seepage. As water flowing through preferential paths can cause streaming potentials and changes in the dam material resistivity and temperature, self-potential (SP) surveys were carried out along with electrical resistivity tomography (ERT) surveys and passive temperature measurements in an existing well through the embankment centered over the perturbed area as defined formerly by MMR technique. The objectives of this work were to demonstrate the effectiveness of SPT to detect the existing water inflow and outflow at dam-water interface as well as the flow path in the dam.

Description of Les Cèdres Embankment Dam and Previous Works

Les Cèdres embankment dam is located south west of Montreal, Canada, (Fig. 1). It was built to guide water from a channel of the St-Lawrence River (Les Cèdres reservoir) to a power plant. The dam separates two reservoirs, Les Cèdres reservoir in the upstream side and St-Timothée reservoir on the downstream side, respectively. St-Timothée reservoir is emptied during the winter and filled during the summer. The top of the dam is flat and is located at an elevation of 42 m. Water levels are at the elevations 40.35 and 35.5 m in Les Cèdres and St-Timothée reservoirs respectively when the latter is filled.

From old plans and geotechnical investigations (Qualitas 2010), it appears that internal structure of the dam (Fig. 2) can be considered as a three layer system composed from top to bottom: heterogeneous material composed from large limestone blocks, gravel, sand, wood, metallic objects, etc., (2) natural clayey silt and (3) non-injected bedrock. The impervious zone is located on the upstream face and is made of clay. Dam height is approximately 12 m and its topography is flat. Our works focus on the western part of the dam where five water outlets were observed on the downstream dam side (on the side facing St-Timothée reservoir). The survey area is located east of the spillway. It is 120 m long starting from spillway and working toward the east.

Geotechnical investigations by drilling and piezocone soundings (Qualitas 2010) show an evidence of water circulation at two places, EO1 and EO2 (Fig. 2). These two water inlets were confirmed by MMR surveys (Fig. 3b) and an additional new inlet, E03, was highlighted (Fig. 2). Optical fiber and thermistor cables were installed in a well through the embankment centered over an infiltration area powered by E02 and E01 water inflow as defined by the MMR survey. The vertical temperature profile versus period of the year (Fig. 3a) shows that most of the changes are caused by advection in the upper 6 m and by conduction in the deeper part. In the other words, there is no evidence of water infiltration.

72   A. Bouchedda et al.

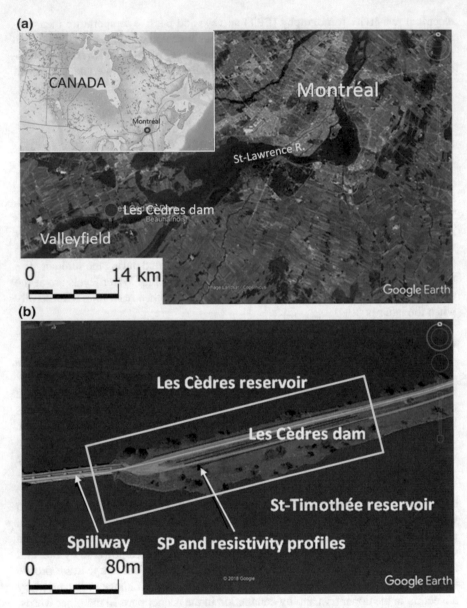

Fig. 1 Les Cèdres embankment dam and survey location. **a** Location of Les Cèdres dam near the
city of Valleyfield (Québec, Canada) 40 km southwest of Montreal, along the St-Lawrence River;
b airborne view of Les Cèdres dam separating Les Cèdres reservoir (north) from St-Timothée
reservoir (south). The SP-ERT survey area is indicated by a yellow rectangular frame

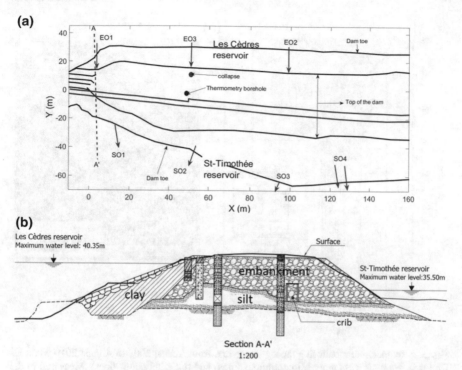

Fig. 2 **a** Plan of view of the dam; observed water outlets and inlets are denoted SO (SO1 to SO4) and EO (EO to EO3), respectively. Red lines delimit the contours of the crib projected at ground surface (the crib is a wooden structure used for transporting embankment material during dam construction); **b** North-South section AA′ showing internal structure of Les Cèdres dam. From top to bottom: (1) heterogeneous material composed from large limestone blocks, gravel, sand, wood, metallic objects, (2) natural clayey silt and (3) non-injected bedrock. The impervious zone made of clay is located on the upstream side. Four geotechnical boreholes are shown that were used to confirm dam construction

Methods

Self-potential Tomography

The Self-potential method is a passive geophysical technique that measures naturally occurring electrical potentials resulting from electro-chemical reactions, temperature gradient, electro-filtration phenomena and other unknown origins (Revil and Jardani 2013). In the case of seepage in embankment dam, we are interested in the potential field along the flow path generated by water flowing through porous media giving rise to a self-potential called streaming potential (Sheffer and Oldenburg 2007). This electro-filtration or electrokinetic phenomenon is due to electrokinetic coupling between the fluid ions and grain capillary wall. In practice, it is very difficult to separate the SP origins. In other words, to characterize water flow, electrofiltra-

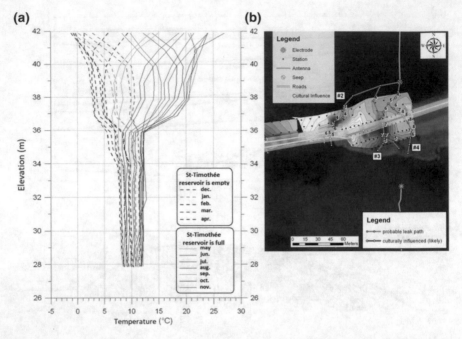

Fig. 3 **a** Temperature monitoring during two years, from August 2008 to August 2010 when St-Timothée reservoir was empty (discontinuous lines) and full (continuous lines). Measured in the thermometry borehole shown in Fig. 2, **b** 2-D MMR map showing interpretation of subsurface water seepage pathways (gray lines) through the dam

tion phenomena should dominate or, at least, other origins should be identified to avoid any misinterpretation. Temperature gradients inside the dams are generally too weak to generate a significant voltage. However, corrosion of buried metallic objects (electrochemical origin) can create large amplitude SP anomalies (≥ 100 mV) (Revil and Jardani 2013).

Self-Potential Tomography (SPT) for leak detection in embankment dam consists in estimating current sources or current densities or hydraulic head distributions caused by water flow through leakage path (Sheffer and Oldenburg 2007). In the following, we will first describe the physics of SP method. Then the forward and inverse problems will be presented. Finally, SP interpretation is discussed.

Physics Principle

When electrokinetic flux is considered, the current density is caused by the sum of convection current, induced by hydraulic head gradient, and the resulting conduction current (Revil and Jardani 2013). The total current density is given by

$$J = J_{conv} + J_{cond} \tag{1}$$

The convection and conduction currents are given by (Sheffer and Oldenburg 2007):

$$J_{conv} = -L\nabla H, \tag{2}$$

$$J_{cond} = -\sigma\nabla v, \tag{3}$$

where L is a coupling coefficient, v the spontaneous polarization electrical potential, H the hydraulic head and σ, the electrical conductivity of the medium.

As the current density is a spatially continuous function and, in the absence of external source (only natural source), we can write:

$$\nabla.J = \nabla.(J_{conv} + J_{cond}) = 0 \tag{4}$$

which gives the following forward problem governing equation for SPT (Sheffer and Oldenburg 2007):

$$\nabla.(\sigma\nabla v) = -\nabla.(L\nabla H) = -\nabla.(J_{conv}) = s, \tag{5}$$

where s can be considered as electrical point sources (Minsley et al. 2007) as in electrical resistivity tomography (ERT); s is a vector of zero values except at current electrode positions where s is equal to the current intensity.

From Eq. (5), the inverse problem can be solved to retrieve hydraulic head or current density or current source distribution. In all cases, the conductivity distribution should be known before SP inversion. In practice, it is obtained from ERT survey carried out in the area under investigation.

Forward Problem

As no analytical solution exists in 3D, the SP forward problem is solved by discretizing Eq. (5); the left term that contains the conductivity and the electrical potential is identical to the ERT forward problem. In our case, we used the finite-volume discretization scheme on a regular grid as described by Sheffer (2007) and Scheffer and Oldenburg (2007). Hence, Eq. (4) can be expressed as

$$A\,v = -D \cdot J_{conv} = s, \tag{6}$$

where A is the discrete version of the Poisson's operator $\nabla.(\sigma\nabla)$.

The matrix A can be written as:

$$\mathbf{A} = \mathbf{D} \cdot \mathbf{M} \cdot \mathbf{G}, \tag{7}$$

with $\mathbf{D}$ the divergence matrix, $\mathbf{M}$ the harmonic mean conductivity matrix and $\mathbf{G}$ the gradient matrix (Sheffer 2007).

The matrix formulation of the SP forward problem can be expressed as follows (Sheffer 2007):

$$\mathbf{v} = \mathbf{P}\mathbf{A}^{-1}\mathbf{s} = \mathbf{P}(\mathbf{D} \cdot \mathbf{M} \cdot \mathbf{G})^{-1} \cdot \mathbf{s} = \mathbf{P}(\mathbf{D} \cdot \mathbf{M} \cdot \mathbf{G})^{-1} \cdot \mathbf{D} \cdot \mathbf{J}_{conv} \tag{8}$$

where $\mathbf{P}$ is interpolation matrix which interpolates the potential from the grid positions to measurement stations.

The Eq. (8) can be written in compact form:

$$\mathbf{v} = \mathbf{K} \cdot \mathbf{m}, \tag{9}$$

where $\mathbf{v}$ is the measured SP data, $\mathbf{K}$ the sensitivity or forward problem matrix. In the case of source inversion: $\mathbf{K} = \mathbf{P}(\mathbf{D} \cdot \mathbf{M} \cdot \mathbf{G})^{-1}$ and $\mathbf{m} = \mathbf{s}$, whereas for current density inversion $\mathbf{K} = \mathbf{P}(\mathbf{D} \cdot \mathbf{M} \cdot \mathbf{G})^{-1}\mathbf{D}$ and $\mathbf{m} = \mathbf{J}_{conv}$.

As the conductivity distribution is known, the SP forward problem is linear. However, as the number of measurements is less than the number of estimated parameters, $\mathbf{K}$ is ill-conditioned. Hence, the inverse problem is non-unique and should be regularized.

Inverse Problem

The inverse problem is solved using a re-weighted iterative regularized least-squares technique (Zhdanov 2015) applied to the minimization of the following model functional

$$\phi(\mathbf{m}) = \|\mathbf{W_d}(\mathbf{v} - \mathbf{Km})\|_2^2 + \beta\|\mathbf{W_c}\mathbf{C} \cdot \mathbf{W_z}(\mathbf{m} - \mathbf{m_{ref}})\|_2^2 \tag{10}$$

where $\mathbf{m}$ is the current density ($\mathbf{m} = (J_x, J_y, J_z)^T$ or the current source ($\mathbf{m} = \mathbf{s}$) model, $\mathbf{m_{ref}}$ is the current density or current source reference model, $\mathbf{d}$ is the measured SP data, $\mathbf{K}$ is the forward modeling matrix and $\mathbf{W_d}$ is the data weighting matrix.

The first term in Eq. (10) is the data fitting function and the second term is the model objective function or stabilizing functional, and β is the regularization parameter. The regularization matrix $\mathbf{C}$ is defined as the combination of the identity matrix $\mathbf{I}$ and the first derivative matrices $\mathbf{D_x}$, $\mathbf{D_y}$ and $\mathbf{D_z}$ in x, y and z directions, respectively. The regularization matrix can be written as:

$$\mathbf{C} = \alpha_x\mathbf{D_x} + \alpha_y\mathbf{D_y} + \alpha_z\mathbf{D_z} + \alpha_s\mathbf{I}, \tag{11}$$

where α_x, α_y and α_z are smoothing weight factors in x, y and z direction, respectively, and α_s is a smallness (or closeness) weight factor. The minimization of the objective function (10) using a Gauss-Newton algorithm results in the following iterative equations:

$$\mathbf{m} = (\mathbf{K}^T \mathbf{W}_d^T \mathbf{W}_d \mathbf{K} + \beta \cdot \mathbf{C}^T \mathbf{C})^{-1} (\mathbf{K}^T \mathbf{W}_d^T \mathbf{W}_d(\mathbf{v}) - \beta \cdot \mathbf{C}^T \mathbf{C}(\mathbf{m}_{ref})) \qquad (12)$$

Potential field approaches such as gravity, magnetic and SP suffer from geometrical decays of their kernel. As consequence, all bodies resulting from the inversion are located near the surface (Li and Oldenburg 1998). In the case of surface SP measurements, Bolève et al. (2009) proposed to include in the inversion system a depth weighting function similar to the one implemented for gravity inversion by Li and Oldenburg (1998).

SP Interpretation

Seepage through an embankment dam, going from the upstream side to the downstream side, will typically create a negative SP anomaly at water inlet and a positive SP anomaly at water outlet (Ogilvy et al. 1969). Hence, SP mapping enables qualitative interpretation to identify problematic zones. However, interference such as buried metallic objects (electrochemical origin) and sharp resistivity contrasts (permeability changes) should be taken into account to avoid any misinterpretation. In other words, negative and positive SP anomalies identification on SP map is not sufficient to locate water inlets and outlets. On the contrary, in SPT the resistivity is firstly estimated by ERT that allows identifying all anomalies related to high resistivity contrasts. As the SP response caused by water flow (electrokinetic origin) and buried objects (electrochemical origin) cannot be separated in practice, it necessary to identify the buried objects to discard the corresponding SP data. For this purpose, we used magnetic and frequency electromagnetic methods (Telford et al. 1995) in our case.

As shown in Eq. (5), SPT depends on hydraulic heads (**H**) or current densities (**J**) or source spatial distributions (**s**). The current density is related to Darcy velocity U (m/s) (Revil and Jardani 2013) by:

$$\mathbf{J}_{conv} = \mathbf{Q}_v \mathbf{U}, \qquad (13)$$

with Q_v (C/m) is the effective excess charge density dragged by the flow of the pore water or more precisely the fraction of the diffuse layer dragged by the water flow. The relationship between the latter and the electrokinetic coupling coefficient L (in V/Pa) is given by (Revil and Jardani 2013)

$$L = -\frac{Q_v k}{\sigma \eta} \qquad (14)$$

where k is the permeability and η the fluid viscosity.

Hence, the determination of Q_v requires permeability measurements.

As current density is a vector quantity, estimated current densities can give us invaluable information on water flow direction, from water inlets to outlets positions, through the dam to map the leakage paths.

Interpretation of current-source models is not intuitive because the relationship between point current sources and water flow cannot be explicitly observed from Eq. (5). However, if we replace (13) in (5), we have

$$\nabla.(\sigma\nabla v) = -\nabla.(Q_vU) = -U\nabla.(Q_v) - Q_v\nabla(U) \qquad (15)$$

Equation (15) shows that a current source is created when $\nabla.(Q_v)$ or $\nabla(U)$ are not 0. Hence, in steady state regime ($\nabla.(U) = 0$), current sources exist when Q_v is spatially variable. As the electrical double layer, responsible for generating the streaming potential, exists only at the mineral-pore water interface and not in the water, there is no excess of charges in the water ($Q_v = 0$). Consequently, water inlet will create negative current source and water outlet will create a positive current source. In heterogeneous dam like Les Cèdres, spatial variation of Q_v will cause current source. In transient regime ($\nabla.(U) \neq 0$), current sources will appear at the positions of permeability changes. The sign of the source will depend on Q_v or k variations.

Electrical Resistivity Tomography

Electrical resistivity tomography (ERT) is a geophysical technique for imaging subsurface resistivity structures from electrical resistivity measurements made at the surface from an array of electrodes. Lateral resolution is controlled by the smallest electrode spacing while penetration is dependent on the largest dipole separation (refer to Loke 2016 for further reading on ERT).

The objective of ERT surveys in this project is twofold: on one hand, to determine the 3-D resistivity distribution of the dam in order to be used in SPT inversion (see Eq. 5); on the other hand, to identify possible leaks zones and determine the internal dam structure (Sjödahl et al. 2008). In fact, the resistivity of material in unsaturated zone decreases in the leak path zones. For example, if we consider Archie's law (see Schön 2015) for unconsolidated saturated material,

$$\rho_{eff} = a\phi^{-n}S^{-m}\rho_w \qquad (16)$$

with a = 0.63, n = 2.15 ϕ = 0.3 and $\rho_w = 48\ \Omega$ m, the effective resistivity ρ_{eff} decreases from 4500 to 402 Ω m, if the saturation S_w increases from 0.3 to 1. The water resistivity ρ_w was measured using a conductivity probe lowered in Les Cèdres reservoir at a depth of 1 m. It is important to note that when clay content increases the effective resistivity of dam material decreases due to surface conductivity (Schön

2015). In practice, knowledge of material in the dam as observed from borehole or construction plan allows separating the two effects.

ERT data were inverted in 3-D using Res3dinv software (Loke 2016). We use robust inversion with the same smoothing factors in all directions. The regularized inverse problem is solved using Gauss-Newton algorithm without approximating the sensitivity matrix.

Magnetic and Frequency Domain Electromagnetic Surveys

The corrosion of buried metal creates SP signal on the order of a few tens of mV that can mask the relatively small signals associated with seepage anomalies. In addition, the forward problem described by Eq. (5) does not take into account this effect. In order to recognize possible buried metallic objects, magnetic and frequency domain electromagnetic measurements were carried out on the SP survey area.

The magnetic data was collected using a highly-sensitive GSMP-35 potassium ground magnetometer-gradiometer (GEM Systems, Canada) in a walking mode using integrated GPS for positioning.

EM in-phase and conductivity data was collected using an EM31-MK2 system (Geonics Ltd, Canada) in vertical dipole mode.

Data Acquisition

Surveys consisting of 2D ERT, self-potential, magnetic (TMI) and magnetic gradient, and EM conductivity profiles were completed at the field site between April and June 2012.

SP measurements were carried out along six 120 m-long profiles named P1, P2, P3, P4, P5 and P6 (Fig. 4). The inter-station spacing was 4 m. All profiles are located on the top of the dam, except P6 which was taken in the water (Les Cèdres reservoir side), parallel to the dam at a distance of about 1 m. SP data were acquired between moving electrodes and a reference electrode (positioned on Les Cèdres dam at 150 m east from survey area). Pb–PbCl$_2$ non-polarizable electrodes (Petiau 2000), manufactured by SDEC (France), were carefully installed to insure good coupling with the ground. SP voltages were measured using a SAS1000 Terrameter (ABEM) by integrating potential on 16.6 ms window for a period 4 s long. The 16.6 ms window was selected to reduce 60 Hz power line interference by stacking. Data was repeatable within 4 mV with standard deviations from 1 to 3 mV.

Electrical resistivity tomography (ERT) profiles were carried out using a Terrameter LS (ABEM, Sweden) with dipole-dipole array and 2 m electrode spacing. In order to reduce the 3-D effect of Les Cèdres dam on ERT inversion, four perpendicular profiles, named T1, T2, T3, T4 were added to P1, P2, P3, P4 and P5 profiles (Fig. 4).

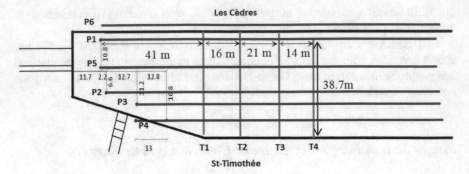

Fig. 4 Sketch of SP and ERT profile location in the survey area located on the top of the dam. SP measurements were carried out along six 120 m long pro-files named P1, P2, P3, P4, P5 and P6. The distance between stations was 4 m. All profiles are located on the top of the dam, except P6 which was taken in the water

Magnetic field and magnetic gradient measurements were collected at 0.5 m intervals along non-regularly spaced profiles. EM31 apparent conductivity data were collected every 1 m along 6 NS lines (L1 to L6) perpendicular to the SP survey lines.

Both magnetic and EM conductivity surveys were limited to the western part of the SP survey area because most spurious SP anomalies assumed to be caused by metallic artefacts were observed in that part.

Results

SP Mapping

SP mapping (Fig. 5) was conducted two months after the completion of St-Timothée reservoir impoundment (June 27–28, 2012). As predicted by SP theory (see interpretation section), all water outlets are associated to positive anomalies. In addition, the location of water inlets, that are theoretically associated to negative anomalies, cannot be well identified from the SP map. In fact, a large elongated negative anomaly can be observed along profile P1 between $X = 0$ m to $X = 90$ m. This anomaly seems to be separated into two anomalies on profiles P5, P2, and P3. It appears with an L shape on the SP map. The minimum value of this anomaly is -30 mV. Note that the SP signal on profile P6 located in Les Cèdres reservoir is positive and not negative as in the profile P1. This is due to the conductivity contrast between water (conductive) and dam material (resistive). The wooden crib (red lines in Fig. 5) seems to have a weak effect on water flow in the dam. As observed, there is no elongated anomaly in the same direction of the crib.

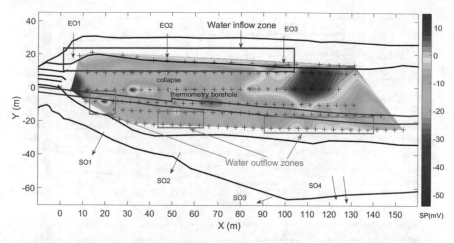

Fig. 5 SP map acquired two months after the completion of St-Timothée reservoir impoundment. Units are in mV. Red arrows are observed water outlet, crosses are SP stations and red lines delimit the contours of the crib projected at ground surface (the crib is a wooden structure used for transporting embankment material during dam construction)

Electrical Resistance Tomography

According to the resistivity model (Fig. 6), the dam structure can be separated into three zones. Zone 1 is located in the upstream dam face and contains the clayey and silty waterproof cover layer. The latter can be easily identified by its low resistivity (<50 Ω m, blue color) between P1 and P5 profiles. However, it appears discontinuous where two resistive (150–600 Ω m) areas can be identified near the reservoir-dam interface. The first one is located between X = 10 m and X = 50 m, and the second one is located between X = 110 m and X = 130 m. Note that the water level in Les Cèdres reservoir is located 0.4–0.8 m below profile P1. These two areas that appear until 4 m deep can be considered as possible zones for water inlets. Zone 2 is located in the center of the dam. It is characterized by high resistivity. It consists of coarse heterogeneous material containing a high proportion of limestone blocks as described in the section: description of Les Cèdres embankment dam and previous works.

Zone 3 is located in the downstream dam face, in contact with the St-Timothée reservoir. At the centre of this zone, a low resistivity area (blue color), extending from X = 50 m to X = 130 m, can be clearly identified. This could be explained by a higher water content caused by water infiltration. In other words, this zone could be considered as the consequence of the three water outlets observed visually between X = 50 m and X = 130 m.

In summary, ERT data allowed us to identify possible infiltration or exfiltration areas. However ERT was not able to resolve if they are water entries or inlets ("sources") or water exits or outlets ("sinks") and if there are one or many sources and their exact locations. Combination in the SPT inversion of the electric resistivity

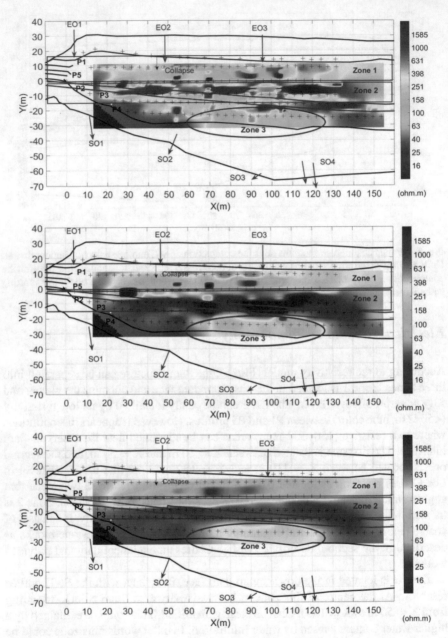

Fig. 6 3D resistivity model from ERT survey. From top to bottom: resistivity depth section at 2.5, 3.4 and 4.4 m respectively. The dam structure can be separated into three zones. Zone 1 contains the clayey and silty waterproof cover layer (<50 Ω m) between P1 and P5 profiles. Note that the water level in Les Cèdres reservoir is located 0.4–0.8 m below profile P1. Zone 2 is characterized by a high-resistivity coarse, heterogeneous material containing limestone blocks. Zone 3, in contact with the St-Timothée reservoir, shows a low resistivity area extending from X = 50 m to X = 130 m, suggesting a higher water content caused by water leakage

model obtained by ERT with the SP data sensitive to water flow should improve the determination of water seepage paths in the dam.

Magnetic and Frequency Domain Electromagnetic Surveys

The magnetic gradient and the apparent electrical conductivity maps are shown in Fig. 7. A strong magnetic anomaly in the area between X = 0 m and X = 20 m, Y = −5 m and Y = 5 m can be clearly identified. It could be associated to an old rail track identified on Hydro-Québec's archive plans. The apparent conductivity map shows two elongated anomalies. The first one corresponds to the effect of the metal-shielded electrical cable that supplies electricity to the electrical spillway gate system. In fact we observed during the survey that the maximum amplitude of the anomaly corresponded to the position of the stakes used for cable identification. The second anomaly that is parallel to the first one could be associated to buried railway track or non-identified metal-shielded electrical cable. It should be noted that the extension of the two anomalies was followed up to about 100 m west of the last profile (L6), which is at 86 m from entrance of the bridge. The measurements were not recorded but we observed the presence of these two anomalies up to about 180 m from the bridge. Before inversion of the SP data, we have eliminated the nine first SP measurements stations of profile P5 that are probably affected by the presence of metallic objects as detected by magnetic survey.

SP Tomography

A 3D model was meshed to invert the measured SP data. Cell dimensions in the region of interest were 1 m in X and Y directions, and from 0.5 m increasing to 2.5 m in Z direction. The 3D model volume in this region was 160 m in X, 70 m in Y and 14.5 m in Z. Conductivities within each cell were constrained to the values obtained from 3D inversion of the ERT data. A depth weighting power of 2 was used. After a few different trials the "best" resulting model of current sources and current densities (lowest misfit and consistent with known outlet locations) was obtained. Figure 8 shows the current density depth section at a depth of 4 m extracted from the 3D model. Contrary to the SP map, now water inflow zones consist of four restricted areas, named ZE1, ZE2, ZE3, and ZE4. ZE1 and ZE2 are characterized by high current density values that would correspond to a large water circulation. In addition, ZE2 corresponds to the water inlet E02 as detected by geotechnical survey, whereas ZE1 is located in front of an area showing ground collapse. ZE3 and ZE4 have low current density amplitude and are considered as potential weak water inlets.

At a depth of 4 m, only the water outlets S04 seem to be present which is consistent with their positions, in the middle of the downstream side. The effect of S01 and

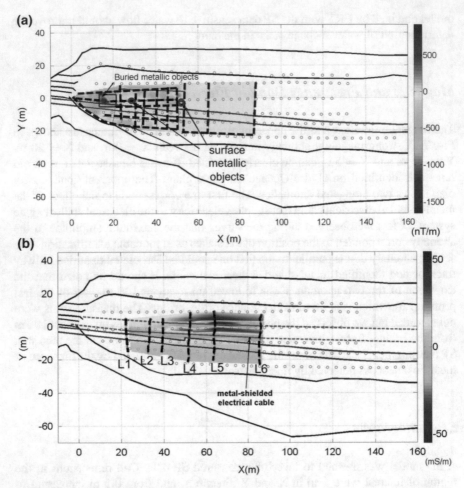

Fig. 7 **a** Magnetic gradient map; units nT/m; measured using magnetic gradiometer in walking mode; metallic objects are located by brown circles and brown rectangle; **b** apparent conductivity map measured using an EM31 in vertical mode along 6 lines (L1 to L6); units mS/m. Black lines are profiles of (**a**) magnetic gradient and (**b**) EM31 conductivity measurements. The plain red lines are projections of the crib at ground surface. Red circle symbols are the actual SP measurements (see Fig. 5). Pink dot lines are buried metal-shielded cables

S03, which are located in the dam toe, can be observed at the 12 m depth section as shown in Fig. 9.

A better idea on water inlet locations and on direction of water flow can be obtained from vertical cross-sections (XZ) of current density and current sources. Figures 10 and 11 show the vertical cross-sections on the upstream side, i.e. at position Y = 10 m. In the current source section (Fig. 10), it can be seen that the sources of the water inlet zones ZE1, ZE2, ZE3 and ZE4 are between 2 and 5 m depth, whereas the current density (Fig. 11) shows that the water flow is mainly downwards.

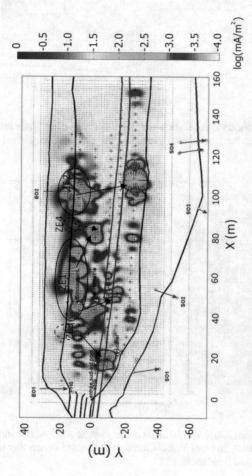

Fig. 8 Horizontal current density cross-section at a depth of 4–4.5 m. Water inlets are denoted with E letter and water outlets are denoted with S letter. Small red arrows are current density vectors (J_x, J_y)

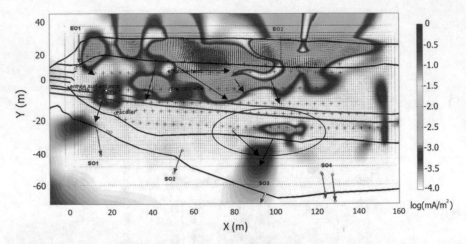

Fig. 9 Horizontal current density cross-section at a depth of 12–14.5 m. Water inlets are denoted with E letter and water outlets are denoted with S letter. Small red arrows are current density vectors (J_x, J_y)

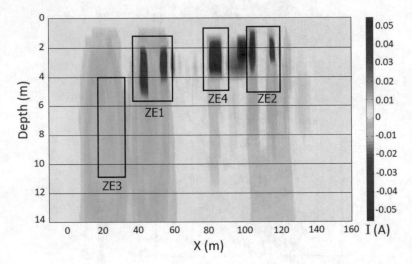

Fig. 10 Vertical cross-section (XZ) at Y = 10 m of current sources obtained from the SPT inversion model. The negative anomalies (in blue) indicate location of seepage connecting water inlets E to water outlets S (see Figs. 8 and 9 for location)

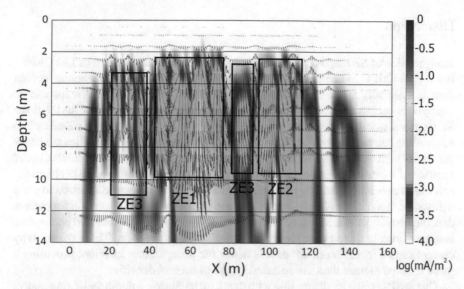

Fig. 11 Vertical cross-section (XZ) at Y = 10 m of current densities obtained from the SPT inversion model Sources displayed in Fig. 10 are partly caused by water flowing downwards

To determine the water flow path in the dam, we used the direction given by the orientation of the current density vectors on the horizontal plane. We tried to follow them from the inlet to the outlet areas. Black arrows on current density sections in Figs. 8 and 9 highlight the interpreted path for each leakage. It is interesting to observe that the water that comes from the main inlet zone ZE1 does not cross the dam in straight line from the upstream side to the downstream side. The water flow seems to circumvent the thermometry borehole and more generally the area delimited by profiles P1 and P5, between X = 50 m and X = 70 m (Fig. 8). This could be explained by the presence of a discontinuous impermeable clay core as shown in electrical resistivity imaging. That would provide an explanation why the effect of water flow was very weak on borehole thermometry measurements as shown before (Fig. 3a).

According to the interpreted current-density model (Figs. 8 and 9), the main outlet zones S03 and S04 are supplied by the water inlet zone ZE2 and, to a lesser extent, by zones ZE1 and ZE4. It can also be observed that ZE3 and ZE1 inlets supply the water for outlets S01 and S02. Finally, it appears that the wooden crib does not control, or only weakly, the water flow when the St-Timothée reservoir is full since negligible current densities are shown parallel to the crib orientation.

Discussion

Interpretation of SP map is critical in the case of heterogeneous dams as Les Cèdres. Ikard et al (2014), when characterizing seepage through an heterogeneous earthen dam in Colorado, USA, show that the interpretation of ERT and SP maps can be ambiguous because the resistivity is not directly related to water flow and that the SP signals distribution is controlled by the resistivity distribution. As SPT takes into account the heterogeneity of resistivity model of the dam, the SP anomalies are better correlated with water flow. Actually SPT inversion for current densities allows a better location of water inlets and outlets. However, as all geophysical inverse problem the solution is non-unique and regularization is needed in the objective function to yield a plausible model. The effect of regularization is more severe in inversion for current densities in comparison to source current because the number of estimated parameters is approximately three times higher in current density inversion. This explains why current source inversion of SP data is better for water inflow location, providing a more detailed picture than the smeared image of current densities.

Due to SP sensitivity decreasing with depth to the source, a depth weighting matrix was introduced in the inversion system to avoid that all the causative sources be found near the surface as demonstrated by Rittgers et al. (2015). This heuristic function is defined as the inverse of depth to the power β, the depth-weighting factor, which is unknown. In our case, we estimated the "best" value from numerical modeling by simulating water inlets at different depths and choosing the best factor that reproduce by inversion the actual location at depth. This factor was found to be about 2 for our dam models. For β larger than 2, the solution overestimate depth and the source is vertically stretched while for β smaller than 2, the solution underestimates depth. Finally, it is important to note that for all potential field methods the deeper is the source the more smeared (low resolution) is the reconstructed image. In other words, the resolution is inversely proportional to the distance between the source and the measurements positions. This is why deeper sources have low amplitude values and they appear more spread than they really are.

Notwithstanding limitations stated above SPT results on Les Cèdres dam are in good agreement with well-known water inflow (on Les Cèdres reservoir side) and outflow positions (on the St-Timothée reservoir). The current densities are in good agreement with expected horizontal and vertical directions of water flow. The crib which we expected would act as an internal channel diverting water parallel to the dam does not appear to affect water flow as evidenced by the mapped current densities. It is possible that the wooden structure was filled, at the end of dam construction, with embankment material having hydraulic properties similar to the dam material.

Water flow in the dam was previously misinterpreted by MMR measurements because the impermeable clay layer is conductive and the electrical current is diverted in as some water infiltration zone. As SP is sensitive to water flow and not only to resistivity contrasts as MMR or ERT, it can better detect seepage in the dam.

Our experiment to map seepage through an embankment dam is similar to the work reported by Bolève et al. (2009). We use ERT data to estimate the distribution of resistivity in the dam. However we do not solve for seepage velocity or flow because it would require to measure some hydrogeological properties of the materials in the dam. Here we limit our interpretation to mapping the water paths or water inlets/outlets by imaging current densities or current sources. We advocate the use of magnetic and EM surveys to help discard spurious SP anomalies of cultural origin.

Conclusion

Geophysical investigations on well-known water seepage in heterogeneous embankment dam, such as Les Cèdres, provide tremendous opportunities for their validation at real scale. In this study, SPT has been successfully used to locate well-known water inflows and outflows. The water flow through the dam appears complex. It is better assessed by SPT in comparison to MMR technique. SPT shows that the permeable zone in the downstream dam side controls the water flow. Magnetic and frequency electromagnetic surveys were useful to discard SP anomalies caused by buried metallic objects and consequently, avoid any water flow infiltration misinterpretation.

Acknowledgements We would like to thank Hydro-Quebec for support for this project on seepage detection and modeling. Financial support has been provided by the Natural Science and Engineering Research Council of Canada and by Hydro-Quebec (CRD Grants Program).

References

Al-Saigh NH, Mohammed ZS, Dahham MS (1994) Detection of water leakage from dams by self-potential method. Eng Geol 37 (2): 115–121.

Bolève A (2013) Internal erosion detection in a concrete-lined canal using a multi-channel self-potential approach. In: Abstract of SAGEEP 2013, Denver, Colorado, 17–21 March 2013.

Bolève A, Vandemeulebrouck J and Grangeon J (2012) Dyke leakage localization and hydraulic permeability estimation through self-potential and hydro-acoustic measurements: Self-potential 'abacus' diagram for hydraulic permeability estimation and uncertainty computation. J appl Geophys 86:17–28.

Bolève A, Revil A, Janod, F, Mattiuzzo JL, Fry JJ (2009) Preferential fluid flow pathways in embankement dams imaged by self-potential tomography. Near Surf Geophys 7(5): 447–462.

Bolève A, Revil A, Janod F, Mattiuzzo JL, Jardani A (2007) Forward modelling and validation of a new formulation to compute self-potential signals associated with ground water flow. Hydrol Earth Syst Sci 11 (5): 1661–1671.

Butler D, Llopis JL, Dobecki TL, Wilt, M, Corwin R, Olhoeft G (1990) Part 2: comprehensive geophysical investigation of an existing dam foundation: engineering geophysics research and development. The Leading Edge 9(9):44–53. https://doi.org/10.1190/1.1439782.

Gex P (1980) Electrofiltration phenomena associated with several dam sites. Bull Soc Vaud Sci Nat 357 (75): 39–50.

Haines BM (1978) The detection of water leakage from dams using streaming potentials. In: Abstract of the SPWLA Nineteenth Annua Logging Symposium, El Paso, Texas, 13–16 June 1978.

Ikard SJ, Revil A, Schmutz M, Karaoulis M, Jardani A and Mooney M (2014) Characterization of focused seepage through an earthfill dam using geoelectrical methods. Groundwater, 52(6), 952–965, https://doi.org/10.1111/gwat.12151.

Jardani A, Revil A, Bolève, A and Dupont JP (2008) 3D inversion of self potential data used to constrain the pattern of groundwater flow in geothermal fields. J Geophys Res 113:B09204. https://doi.org/10.1029/2007jb005302.

Kofoed, V, Jessop M, Wallace M, and Qian W (2011) Unique applications of MMR to track preferential groundwater flow paths in dams, mines, environmental sites, and leach fields. The Leading Edge 30(2):192–204. https://doi.org/10.1190/1.3555330.

Li Y and Oldenburg DW (1998) 3-D inversion of gravity data. Geophysics 63: 109–119.

Loke MH (2016) Lecture notes on 2D & 3D electrical imaging surveys. Geotomo Software. Available via DIALOG: http://www.geotomosoft.com/downloads.php. Accessed 01 Dec 2016.

Minsley B, Sogade J and Morgan F (2007) Three-dimensional source inversion of self-potential data, J. Geophys. Res., 112, B02202, https://doi.org/10.1029/2006jb004262.

Ogilvy AA, Ayed, MA, Bogoslovsky, VA (1969) Geophysical studies of water leakages from reservoirs. Geophys Prosp 17:36–62.

Panthulu TV, Krishnaiah C, Shirke JM (2001) Detection of seepage paths in earth dams using self-potential and electrical resistivity methods. Eng Geol 59 (3–4): 281–295.

Petiau G (2000) Second Generation of Lead-lead Chloride Electrodes for Geophysical Applications. Pure appl. Geophys., 157:357–382.

Qualitas (2010) Geotechnical investigations, drilling and piezocone soundings on Les Cèdres dam. Report presented to Beauharnois Gatineau direction, Hydro-Québec, 2010, 63p. (In French).

Revil A, Jardani A (2013) The Self-potential Method: Theory and Applications in Environmental Geosciences, Cambridge University Press.

Rittgers JB, Revil A, Planès T, Mooney MA and Koelewijn AR (2015) 4D imaging of seepage in earthen embankments with time-lapse inversion of self-potential data constrained by acoustic emissions localization. Geophysical Journal International, 200, 758–772, https://doi.org/10.1093/gji/ggu432.

Rozycki A, Fonticiella JMR, Cuadra A (2006) Detection and evaluation of horizontal fractures in earth dams using self-potential method. Eng Geol 82(3): 145–153.

Rozycki A (2009) Evaluation of the streaming potential effect of piping phenomena using a finite cylinder model. Eng Geol 104 (1–2): 98–108.

Schön J (2015) Physical properties of rock; Fundamentals and principles of petrophysics; 2nd edition. Elsevier, Amsterdam.

Sjödahl P, Dahlin T, Johansson S, LokeMH (2008) Resistivity monitoring for leakage and internal erosion detection at Hällby embankment dam. J of App Geophysics, 65 (3–4):155–164.

Sheffer M R (2002) Response of the self-potential method to changing seepage conditions in embankments dams. M.Sc. Thesis, Dept of Civil Eng, University of British Columbia.

Sheffer M R (2007) Forward modelling and inversion of streaming potential for the interpretation of hydraulic conditions from self-potential data. Ph.D thesis; University of British Columbia.

Sheffer M R, Oldenburg D W (2007) Three-dimensional modelling of streaming potential. Geophys J Inter, 169 (3): 839–848.

Smith M, Côté A, Noël P, Babin D (2009) Characterizing seepage at the junction of two embankment dams. Abstract of the Canadian Dam Association. Available via DIALOG:http://www.cda.ca/proceedings%20datafiles/2009/2009-9b-01.pdf.

Telford, W.M, Geldart, L.P and R.E Sheriff (1995) *Applied Geophysics*, 2nd Edition, Cambridge University Press.

Zhdanov M (2015) Inverse Theory and Applications in Geophysics, 2nd Edition, Elsevier.

Optical Fiber Sensors for Dam and Levee Monitoring and Damage Detection

Daniele Inaudi

Abstract Optical fiber sensors can be used advantageously for monitoring dams and levees and to detect and localize damage in them. This technology is relatively new, but in the last 20 years numerous applications have been successfully carried out in dams, dykes and levees worldwide. There are two main usage scenarios for optical fiber sensing technology. Some sensors replace conventional, e.g. vibrating wire, sensors with equivalent optical versions. In this case the main benefits come from their immunity to electromagnetic interference—such as lightning strikes or power lines—and the possibility to use cables of up to several km long to connect the sensors and the readout units. The second main application of optical fiber sensing relies on the use of distributed sensors. Those sensors can measure strain and temperature every meter along a cable that can reach several km in length. This enables detecting and localizing undesired events such as leaks or cracks that produce a local change of strain or temperature. In this chapter we will introduce the different optical fiber sensing technologies and corresponding sensors. Several applications example will illustrate the abovementioned use scenarios.

Introduction

Earthen embankments including levees, tailings dams, and earthen dams present many challenging problems for civil engineers, particularly in the verification of their structural integrity and capacity, operation and maintenance (O&M), inspection and safety. The large size, age and uncertainty of material properties in these sometimes mammoth structures, all combine to present a difficult array of parameters for the dam and levee professionals to navigate when analyzing a new or existing levee or dam.

To make things more difficult, there are an ever growing number of assets and lives these structures protect downstream or in the "flood plain," and more and more emphasis is being placed on the vulnerability of these structures. In the wake of

D. Inaudi (✉)
SMARTEC SA, Manno, Switzerland
e-mail: daniele.inaudi@smartec.ch

© Springer Nature Switzerland AG 2019
J. Lorenzo and W. Doll (eds.), *Levees and Dams*,
https://doi.org/10.1007/978-3-030-27367-5_5

flood disasters associated with Hurricane Katrina and others, a complex regulatory environment has emerged; requiring engineers to certify structural and geotechnical fortitude, and levee and dam asset owners and engineers are therefore facing new challenges.

Levees have many different failure modes (Sills et al. 2008). Many of the most common failure modes have indicators (erosion, seepage and/or settlement) that are often difficult to detect with the human eye during a superficial inspection. Relying on visual inspection alone, weaknesses in these structures can remain undetected, until a failure occurs during a storm or surge event. Visual inspection and surveying are vital parts of any levee management program, and neither can or should be replaced; however, current inspection practices have some limitations:

- Levees and dams are too large in scale (many miles long, very wide, very tall) to thoroughly inspect visually and survey consistently.
- Inspections and surveys can be spaced by several months or even years
- Differential settlement, structural weakness, and warning signs can be nearly impossible to detect with visual inspection
- Water, vegetation and other obstruction can limit the surface that can be visually inspected.
- Many structural failures originate underground with no initial surface expression
- Many "warning signs" are not intuitively obvious and difficult to detect even by the most well trained inspectors.

Recent advances in instrumentation technologies and applications are providing new ways for the civil engineer to examine these structures, and offer a set of monitoring tools previously thought impossible. Distributed fiber optic technologies (Inaudi and Glisic 2006) are among those emerging technologies and enable sensing over the whole length or volume of a dam or levee. Those distributed sensors are able to detect and localize defects and damage occurring anywhere along the levee or dam, providing a new and cost effective way of monitoring these structures.

Optical fiber sensors, when used as point sensors to measure quantities such as pore water pressure, strain, displacement, or temperature, present significant advantages over conventional electrical sensors in terms of long-term reliability and insensitivity to external perturbations, such as lightning strikes or proximity to power lines (Udd 2011).

In this chapter, we will review commonly available fiber optic sensing technologies and explore their applications to the monitoring of dams, levees, and other geotechnical applications.

Fiber Optic Point Sensors

From many points of view, fiber optic sensors are the ideal transducers for structural monitoring (Glisic and Inaudi 2008). Being durable, stable and insensitive to external perturbations, they are especially useful for long-term health assessment of civil structures and geostructures. Many different fiber optic sensor technologies exist and offer a wide range of performance and applicability, as will be detailed in the following paragraphs. Fiber optic sensors often offer measurement performance similar to those of the corresponding conventional sensors, such as vibrating wire or electrical sensors. This technology can however offer other advantages in terms of reliability and accuracy when operation in harsh environments or in the presence of electromagnetic disturbance.

Finally, distributed fiber sensors offer new exciting capabilities that have no parallel in conventional sensors.

There exists a great variety of fiber optic sensors (FOS) for structural and geotechnical monitoring. In this overview we will concentrate on those that have reached a level of maturity, allowing a routine use for a large number of structural monitoring applications. Figure 1 illustrates the four main types of fiber optic sensors (Table 1).

The greatest advantages of the FOS are derived from the characteristics of the optical fiber itself that is either used as a link between the sensor and the signal conditioner, or becomes the sensor itself in the case of long-gauge and distributed sensors. In almost all FOS applications, the optical fiber is a thin glass fiber that is protected mechanically by a polymer coating (or a metal coating in extreme cases) and further protected by a multi-layer cable structure designed to shield the fiber from the environment where it will be installed. Since glass is an inert material very resistant to almost all chemicals, even at extreme temperatures, it is ideal for use in harsh environments such as those encountered in geotechnical applications. Chemical

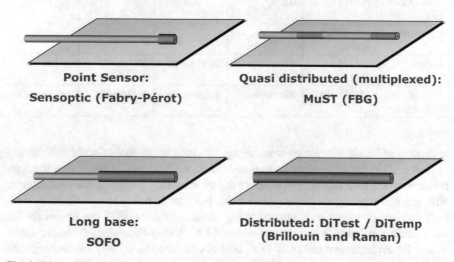

Point Sensor:
Sensoptic (Fabry-Pérot)

Quasi distributed (multiplexed):
MuST (FBG)

Long base:
SOFO

Distributed: DiTest / DiTemp
(Brillouin and Raman)

Fig. 1 Main types of optical fiber sensors: point sensors have a single measurement point at the end of the fiber optic connection cable, similarly to most electrical sensors. They measure a quantity in a small area of typically less than 1 cm length. Multiplexed sensors allow the measurement at multiple points, spaced by a few cm or several meters, along a single fiber line. Long-base sensors integrate the measurement over a long measurement base (typically 1 m or more). They are also known as long-gage sensors. Distributed sensors are able to sense at any point along a single fiber line, typically every meter over many kilometers of length

Table 1 Illustrates the main types of sensors and their characteristics

	Fabry-Perot interferometric	Fiber Bragg gratings	SOFO interferometric	Raman scattering	Brillouin scattering
Type of sensor	Point	Point	Long-gauge (integral strain)	Distributed	Distributed
Measurable parameters	Strain Temperature Pressure Displacement	Strain Temperature Acceleration	Deformation Strain	Temperature	Strain Temperature
Multiplexing (several sensors on same fiber)	No	Yes	No	Fully distributed	Fully distributed
Number of measurement points in one fiber	1	10–30	1	30,000	50,000
Typical accuracy					
Strain	1 $\mu\varepsilon$	1 $\mu\varepsilon$	1 $\mu\varepsilon$	0.1 °C	10 $\mu\varepsilon$
Deformation	10 μm	1 μm	1 μm		0.2 °C
Temperature	0.1 °C	0.1 °C			
Pressure	0.25% full scale				
Range			20 m gauge length	30 km	50 km
Type of optical fiber	Multimode	Singlemode	Singlemode	Multimode	Singlemode

resistance and durability are great advantages for long term reliable health monitoring of civil engineering structures (Inaudi and Glisic 2006; Inaudi 2004). Since the light confined to the core of the optical fibers used for sensing purposes does not interact with any surrounding electromagnetic field, FOS are immune to interference. With such unique advantage over sensors using electrical cables, FOS are obviously the ideal sensing solution when the presence of EM, Radio Frequency or Microwaves cannot be avoided. For instance, FOS will not be affected by any electromagnetic field or electrical currents generated by lightning hitting a monitored dam, nor from the interference produced by a subway train running near a monitored zone. FOS are intrinsically safe and naturally explosion-proof, making them particularly suitable for monitoring applications of risky structures such as gas pipelines, coal mines or chemical plants. But the greatest and most exclusive advantage of such sensors is their ability to offer long range distributed sensing capabilities.

The main disadvantages of fiber optic sensors include the higher cost and complexity of the readout systems and the need to specialized equipment to splice or repair optical fibers.

Let's now review some of the most common optical fiber sensors used for structural and geotechnical monitoring.

Piezometers and Other Point Sensors

Fabry-Pérot Interferometric sensors (Pinet 2009) are typical examples of point sensors and have a single measurement point at the end of the fiber optic connection cable. An extrinsic Fabry-Pérot Interferometer consists of two partially mirrored optical fibers facing each other, but leaving an air cavity of a few microns between them, as shown in Fig. 2. When light is coupled into one of the fibers, a back-reflected interference signal is obtained. This is due to the reflection of the incoming light on the two mirrors. This interference can be demodulated using coherent or low-coherence techniques to reconstruct the changes in the fiber spacing allowing a precise measurement of the gap width, with nanometer accuracy. Since the two fibers are attached to the capillary tube near its two extremities (with a typical spacing of 10 mm), the gap change will correspond to the average strain variation between the two attachment points shown in Fig. 2.

Many sensors based on this principle are currently available for geotechnical monitoring, including weldable and embedded strain gauges, temperature sensors, and displacement sensors. Examples are shown in Fig. 3.

A particularly interesting sensor type for the monitoring of Dams and levees is the Fabry-Perot optical piezometer. In this type of sensor, the second fiber is replaced by a membrane that deforms due to the applied external pressure. This enables piezometers that are externally identical to their conventional vibrating-wire equivalents, as shown in Fig. 4 left, but also miniature sensors adapted for installation in small ducts or in geotextiles (Rodrigues et al. 2010) as shown in Fig. 4 right. Those sensors can be used advantageously in applications where frequent lightning strikes of very long cables limit the use of electrical and vibrating wire sensors.

Fig. 2 Functional principle of a Fabry-Perot sensor

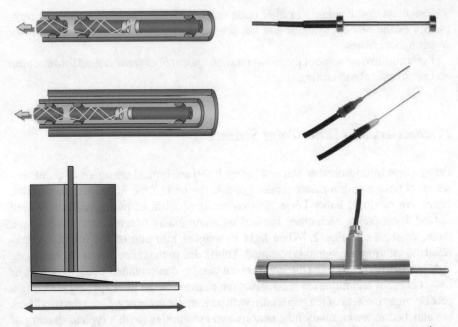

Fig. 3 Examples of strain, temperature and displacement sensors based on Fabry-Perot sensing

Fig. 4 Fabry-Perot piezometer with conventional (left) and miniature (right) packaging

Long-Base Deformation Sensors

The SOFO (French acronym for Structural Monitoring with Optical Fibers) Interferometric sensors are long-base sensors, integrating the measurement over a long measurement base that can reach 10 m or more (Inaudi et al. 1994). Long-base fiber optic displacement sensors offer a resolution in the micrometer range and excellent long-term stability, allowing measurements over more than ten years with micrometer accuracy (Glišić et al. 2005). The measurement setup uses low-coherence interferometry to measure the length difference between two optical fibers installed on the structure to be monitored (Fig. 5), by embedding in concrete or surface mounting.

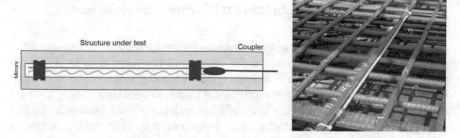

Fig. 5 Functional principle of SOFO sensors and sensor installation in concrete

The measurement fiber is pre-tensioned and mechanically coupled to the structure at two anchor points in order to follow its deformations, while the reference fiber is free and acts as temperature reference. Both fibers are installed inside the same plastic pipe and the gage length can be chosen between 200 mm and 10 m. The readout unit, shown in Fig. 6, measures the length difference between the measurement fiber and the reference fiber with micrometer accuracy.

Those sensors are particularly adapted when an accurate and long-term stable deformation measurement is required. They are particularly useful when installed inside or on the surface of inhomogeneous materials such as concrete, rock, soil, masonry or timber. Thanks to their long gauge length, those sensors average the strain recording and make their reading less sensitive to local material inhomogeneity or defects. Examples of application include the measurement of internal deformations of dams, tailing dams and levees, as well as deformation monitoring of walls, piles and other retaining structures (Glisic et al. 2002). These sensors have been installed inside ground anchors for long-term deformation monitoring (Inaudi and Casanova 2000).

Fig. 6 Example of SOFO
readout unit

Fiber Optic Distributed Strain and Temperature Sensors

Even before the optical fiber sensors described above even existed, many different types of discrete instrumentation for dams have been available to civil engineers. These sensors incorporate different types of sensing technologies an, including piezometers, inclinometers, and settlement plates. Levees traditionally have not included sensors, but occasionally have relied on traditional point sensors for limited data gathering at a specific location or to investigate a specific defect (Dunnicliff 1993). For such extended structures, deciding where to install point sensors becomes a challenge in itself. Placing traditional, discrete sensors made defect detection and localization highly improbable, and was seldom considered a cost effective method in comparison to regular visual inspection.

Through advances in Structural Health Monitoring (SHM) and civil engineering instrumentation, new monitoring technologies and applications have emerged that are complementing the traditional methods engineers use to evaluate and inspect levees and earthen dams. The fiber optic distributed strain/temperature monitoring systems can provide actionable operation and maintenance information, provide powerful inspection and assurance tools, and provide warning to engineers and asset managers about failures before they occur, increasing safety for all dam and levee stakeholders. Now engineers have tools to seamlessly monitor levee segments of lengths up to 30 km, with no breaks or gaps, allowing 100% coverage (see Fig. 7). These distributed sensors allow engineers to both localize and quantify movements and leaks (Inaudi and Glisic 2005). In addition, earthen tailings dams can be hundreds of feet high, and can similarly achieve full same coverage.

A distributed fiber optic monitoring system consists of one or more unique sensor cables (fiber optic) and one readout device. The area of coverage can be up to 30 continuous kilometers in length with one system. The sensor cables can be deployed either during construction or after construction and is possible to retrofit existing structures by cable plowing or trenching. The readout system can monitor strain and temperature along the entire length of cable, and is able to detect the following failure modes (Inaudi and Church 2011):

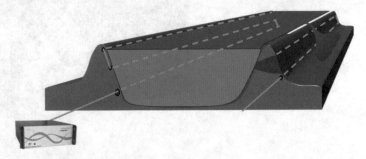

Fig. 7 Use of a single distributed sensor to monitor the strain and temperature along the full length of a levee at multiple levels

- Structural movement or failure,
- Overtopping,
- Under-seepage,
- Through-seepage,
- Piping (internal erosion),
- External erosion,
- Differential settlement,
- Landslides.

Unlike electrical sensors and localized fiber optic sensors, distributed sensors offer the unique characteristic of being able to measure physical parameters, in particular strain and temperature, along their entire length, allowing the measurement of thousands of points from a single readout unit.

The most developed technologies of distributed fiber optic sensors are based on Raman and Brillouin scattering. Both systems make use of a nonlinear interaction between the light and the glass material of which the fiber is made. If an intense light at a known wavelength is injected into a fiber, a very small amount of it is scattered back from every location along the fiber itself. Besides the original wavelength (called the Rayleigh component), the scattered light contains components at wavelengths that are higher and lower than the original signal (called the Raman and Brillouin components). These shifted components contain information on the local properties of the fiber, in particular its strain and temperature at the location where the scattering occurred. Figure 8 shows the main scattered wavelengths components for a standard optical fiber and how they can be sued for sensing purposes.

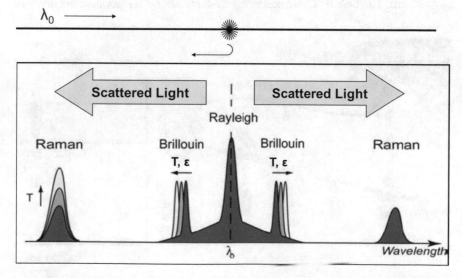

Fig. 8 Wavelength components of scattered light in silica fibers, Raman and Brillouin peak location and intensity are temperature and strain dependent and can be sued for sensing those parameters

If λ_0 is the wavelength of the original signal generated by the readout unit, the scattered components appear at both higher and lower wavelengths. The two Raman peaks are symmetric to the original wavelength. Their position is fixed, but the intensity of the peak at lower wavelength is temperature dependent, while the intensity of the one at higher wavelength is unaffected by temperature changes. Measuring the intensity ratio between the two Raman peaks therefore yields the local temperature in the fiber section where the scattering occurred.

The two Brillouin peaks are also located symmetrically at the same distance from the original wavelength. Their position relative to λ_0 is however proportional to the local temperature and strain changes in the fiber section. Brillouin scattering is the result of the interaction between optical and ultrasound waves in optical fibers. The Brillouin wavelength shift is proportional to the acoustic velocity in the fiber which is related to its density. Since the density depends on the strain and the temperature of the optical fiber, we can use the Brillouin shift to measure those parameters.

When light pulses are used to interrogate the fiber it becomes possible, using a technique similar to RADAR, to discriminate different points along the sensing fiber through the different time-of-flight of the scattered light. Combining the radar technique and the spectral analysis of the returned light one can obtain the complete profile of strain or temperature along the fiber. Typically it is possible to use a fiber with a length of up to 30 km and obtain strain and temperature readings every meter, as depicted in Fig. 9. In the professional jargon, this is referred to as a distributed sensing system with a range of 30 km and a spatial resolution of 1 m.

Systems based on Raman scattering typically exhibit temperature accuracy of the order of ±0.1 °C and a spatial resolution of 1 m over a measurement range up to 30 km. The best Brillouin scattering systems offer a temperature accuracy of

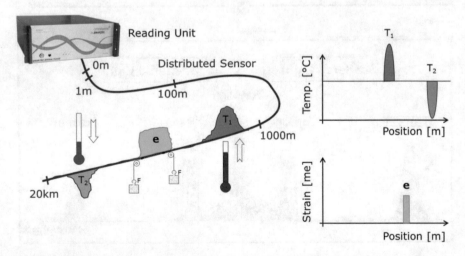

Fig. 9 Example of distributed strain and temperature sensing setup. If the cable is subject to temperature changes (red and blue zones) or strain (green zone) the instrument is able to identify and localize those events as illustrated in the corresponding graphs

± 0.1 °C, a strain accuracy of ±20 microstrain and a measurement range of 50 km, with a spatial resolution of 1 m. The readout units are portable and can be used for field applications.

The optical fibers are only 1/8 of a millimeter in diameter and are therefore difficult to handle and relatively fragile. For practical uses, it is therefore necessary to package them in a larger cable, much like copper conductors are incorporated in an electrical cable. Since the Brillouin frequency shift depends on both the local strain and temperature of the fiber, the sensor set-up will determine the actual response of the sensor. For measuring temperatures it is necessary to use a cable designed to shield the optical fibers from an elongation of the cable (Inaudi and Glisic 2005). The fiber will therefore remain in its unstrained state and the frequency shifts can be unambiguously assigned to temperature variations. Measuring distributed strains also requires a specially designed sensor. A mechanical coupling between the sensor and the host structure along the whole length of the fiber has to be guaranteed. Since the system is sensitive to both strain and temperature variations, it is also necessary to install a reference temperature sensing fiber along the strain sensor, so that the effect of temperature can be removed from the strain reading. Special cables (Fig. 10), containing both free and coupled fibers allow a simultaneous reading of strain and temperature.

Applications to Dam and Dykes Leak/Seepage Detection and Temperature Monitoring

The use of distributed temperature sensing is of particular interest for the monitoring of dams, dykes and levees. Leaks and seepage can be detected and localized by observing temperatures anomalies along the sensing cable. It is therefore important to place the sensing cable at those locations in the levee or dyke cross-section where such undesired events might occur.

The following examples illustrate a few typical applications of distributed temperature sensing for leak and seepage detection.

Water reservoirs in Spain

Type of Structure: Water reservoir with plastic membrane
Monitoring aim: Detect leaks through membrane and perimeter levee
Fiber Optic sensing technology employed: Distributed Raman temperature sensing with heated cable

A water reservoir in Spain consists in a square levee structure and an inner synthetic membrane. This large structure is used to store freshwater and potentially presents a risk of internal erosion if the watertight membrane were to leak. A distributed temperature monitoring cable was installed in the sand below the membrane at two levels around the levee perimeter as shown in Fig. 11 and interrogated with the system depicted in Fig. 12. To enhance the detection capability of the system,

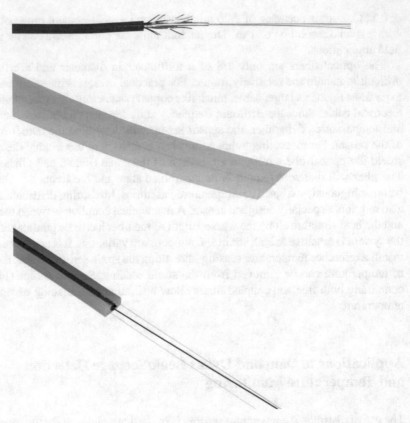

Fig. 10 Example of distributed sensing cables to monitor temperature, strain and combined strain/temperature

Fig. 11 Water reservoir (Spain) after laying waterproof membrane. The sensing cables are installed at the bottom and at mid-height around the circumference of the levee

Fig. 12 Distributed temperature sensing system and ancillary equipment installed at pumping station

a hybrid cable containing both optical fibers and copper wires was used. By circulating an electric current in the copper wires, it is possible to slightly increase the temperature of the cable itself and the surrounding ground. Since dry sand has lower thermal capacity than wet sand, for the same heating power it will increase its temperature more rapidly. By observing the locations along the cable where the temperature increases more slowly, it is possible to identify potential leak locations.

Nam Gum Dam in Laos

Type of Structure: Concrete Face Rockfill Dam
Monitoring aim: Detect leaks through concrete face along the plinth
Fiber Optic sensing technology employed: Distributed Raman temperature sensing

Nam Ngum Reservoir is the largest water impoundment in Laos; it was built in 1971 as a result of the construction of the first dam across the Nam Ngum River. The reservoir was designed primarily for the production of hydro-electric power and flood control. The Nam Ngum 2 Hydroelectric Power Project is located approximately 35 km upstream of the existing Nam Ngum 1 dam, about 90 km from Vientiane, on the Nam Ngum River, which is one of the major tributaries to the Mekong River. The project, with an installed capacity of 615 MW is being built to produce energy for the Thai electricity grid and for local consumption. The dam is 181 m high and it can produce 2300 kW of electric power per hour.

The main aim of the installed instrumentation is to monitor the seepage at the foundation plinth level with an active detection system, using the heated cable method. A total of 2 independent armored sensing cables (approx. 900 m each) have been integrated in the filter zone by surface installation as shown in Fig. 13. The DTS reading unit with 4 channels is located in the dam control station with the aim of measuring the temperature profiles of the 2 sensing cables. Thanks to the customized visualization software it's possible to follow in real time any variation in the temperature profiles (Fig. 14) of the two sensing cable, launching a warning in case of seepage or leakage.

Luzzone Dam in Switzerland

Type of Structure: Concrete Arch Dam
Monitoring aim: Monitor temperature evolution during concrete setting
Fiber Optic sensing technology employed: Distributed Brillouin temperature sensing

In 1997 the Luzzone Dam in Switzerland was raised by 18 m to increase the capacity of the reservoir. Adding fresh concrete on top of existing concrete presented challenges related to differential shrinkage and temperature evolution during

Fig. 13 Distributed temperature sensing cable installation along the plinth dam of the Nam Ngum Dam, Laos

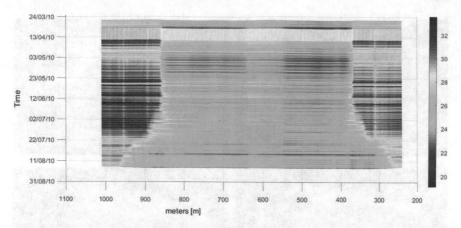

Fig. 14 Temperature evolution along the dam plinth during impounding. The horizontal axis represents position along the plinth, while the vertical axis represents time, with older dates on top. The color represents the recorded temperature in °C. The blue area, representing lower temperatures, expands outwards as the water level increases, indicating that the recorded temperature is mostly influenced by the water temperature and no longer sensitive to air temperature evolution

setting. The new concrete was instrumented with long-gage fiber optic sensors to monitor its shrinkage deformation and to evaluate the interaction between the newly added concrete and the old one (see Fig. 15) (Inaudi et al. 1998). Additionally, a distributed temperature sensor was installed in one of the largest concrete blocks to monitor the temperature evolution during concrete setting. In Fig. 16 it is possible to observe the temperature distribution within the concrete (Thévenaz 1999). The concrete block warms up rather uniformly, but during the cool-down phase, that can last several months, significant temperature gradients of up to 35 °C were observed. Those gradients can lead to concrete cracking.

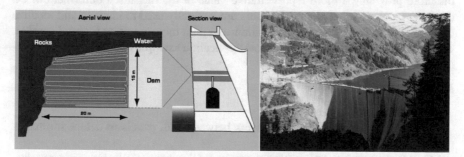

Fig. 15 Luzzone (Switzerland) dam raising, cable location and picture during work

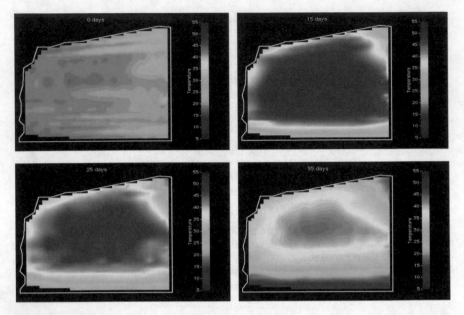

Fig. 16 Evolution of internal temperature in °C in Luzzone dam during concrete setting and after 15, 25 and 55 days

Application to Dams and Levee Deformation Monitoring

Local and distributed optical fiber strain sensors have found several useful applications in dam, dyke and levee monitoring. They are deployed to monitor the deformations of the dam body and its foundations. Long-gauge deformations sensors can be used to monitor large-scale deformations, averaging strain over several meters and therefore providing a more representative value of the measured deformation. On the other hand, distributed strain sensors are generally used to identify and localize damage such as cracks, settlements, abnormal joint movements or internal erosion. The following paragraphs provide a few application examples of those sensing technologies.

I Wall Levee in USA

Type of Structure: I levee wall
Monitoring aim: Monitor movements between wall panels to detect anomalies and impending panel failure
Fiber Optic sensing technology employed: Distributed Brillouin strain sensing

The iLevees project "Intelligent Flood Protection Monitoring Warning and Response Systems", in the state of Louisiana, has the goal of providing an alerting and monitoring system capable of preventing early stage failure, both in terms of ground instability and seepage (Wang et al. 2014). The motivation for the monitoring system is to improve safety awareness, to provide relevant information about levees' status and conditions, before, during and after floods, and to avoid the tragic events like the ones that occurred following Hurricane Katrina in 2005. The use of distributed fiber optic sensing helps in overcoming the issue of optimal sensor location allowing full structure coverage. Continuous long-term monitoring during the complete levee lifetime allows for the collection of data that can improve our general knowledge of these structures, with unquestionable benefits in future levee designs, operation and maintenance. To demonstrate different sensing technologies, a number of test sections have been instrumented; including I-wall and T-wall sections instrumented with distributes strain and temperature sensors. Figure 17 shows the installation of a distributed strain and temperature sensing cables in the levee foundations and on top of the I-wall and T-wall sections. These sensors allow the detection and localization of events such as levee failure onset, seepage, tunneling, and formation of cracks in wall sections or abnormal joint movements.

An example of calculated deformation on the sensor placed on the top of the I-wall section is presented in Fig. 18. Deformation is plotted as a function of position along the wall and as a function of time. In the plot it is possible to observe the daily expansion-contraction cycles of the wall due to temperature fluctuations. It is also possible to localize the expansion joints along the levee wall that shows different behavior. In case of an event along the levee section, a localized deformation peak will appear in the visualization software and would automatically trip an alarm.

Earthen Levee in Netherlands

Type of Structure: Earthen levee
Monitoring aim: Detect early signs of levee failure on full-scale levee test section
Fiber Optic sensing technology employed: Distributed Brillouin strain sensing

Fig. 17 Installation of distributed strain and temperature sensing cables in the levee foundations and on top of the I-wall and T-wall sections in New Orleans, Louisiana, USA

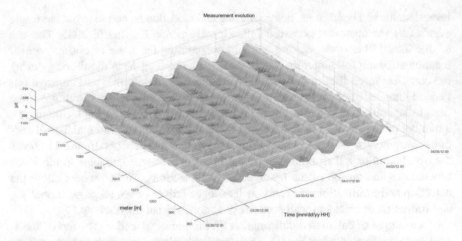

Fig. 18 Strain evolution as a function of time and position along a levee wall section

The IJkdijk was a collaborative project in the Netherlands to test dikes and to develop several existing sensor network technologies for dikes early warning systems. In 2008 a full-scale section of dike was constructed and destroyed in the Macrostability Experiment. The dike section (see Fig. 19) was roughly 100 m long, 30 m wide and 6 m high and consisted of a core of white sand and a shell of clay. A drainage system was placed at the bottom of the sand core, allowing the increase of the internal water pore pressure. Containers were placed on top of the dike, eventually to be filled with water to increase the normal load on top to reach the failure. Among many other sensing technologies, geotextile sensing strips embedding several distributed fiber optic strain cables were installed at four levels along the levee section. Those geotextile sensing strips were able to detect and localize the upcoming failure

Fig. 19 Levee section under test, Netherland. Courtesy of Deltares

zone up to 42 h before the final failure of the levee. Figure 20 shows the evolution of strain along the sensing strips at the 4 different heights (Artières et al. 2010).

River dam in Latvia

Type of Structure: River dam and hydropower plant
Monitoring aim: Detect and localize leaks across bitumen joint
Fiber Optic sensing technology employed: Distributed Brillouin strain sensing

Plavinu hes is a dam belongs to the complex of three most important hydropower stations on the Daugava River in Latvia (see Fig. 21). In terms of capacity this is the largest hydropower plant in Latvia and is considered to be the third level of the Daugavas hydroelectric cascade. It was constructed in 1960, 107 km distant from the firth of Daugava and is unique in terms of its construction. For the first time in the history of hydro-construction practice; a hydropower plant was built on clay-sand and sand-clay foundations with a maximum pressure limit of 40 m. The power plant building is merged with a water spillway. The entire building complex is extremely

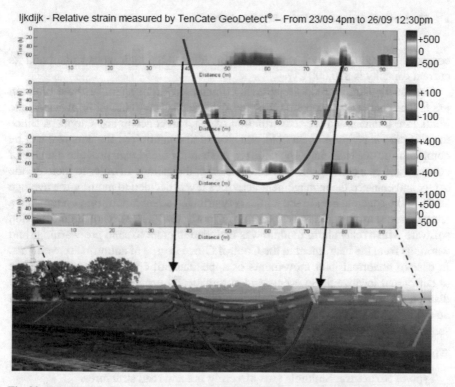

Fig. 20 Distributed strain evolution along four lines of sensing cables and. The four graphs represent the measured strain as a function of position (horizontal axis) and time (vertical axis). Red color indicates higher tensional strain, blue color indicates compression strain. It is possible to observe how the locations of the strain peaks correspond to the boundary zones of the collapsing area in the picture taken after failure. Courtesy of TenCate Geosynthetics

Fig. 21 Aerial photograph of Plavinu hes dam, Latvia

compact. There are ten hydro aggregates installed at the hydropower plant and its current capacity is 870,000 kW.

One of the dam inspection galleries coincides with a system of three bitumen joints that connect two separate blocks of the dam. Due to abrasion of water, the joints lose bitumen and the redistribution of loads in concrete arms appears. Since the structure is nearly 40 years old, the structural condition of the concrete can be compromised due to ageing. Thus, the redistribution of loads can provoke damage of concrete and as a consequence the inundation of the gallery. In order to increase the safety and enhance the management activities it was decided to monitor the average strain in the concrete next to the joints. A distributed strain monitoring system and a temperature sensing cable is used for this purpose (see Fig. 22). Threshold detection software with a relay alarm module was installed in order to send pre-warnings and warnings from the instrument to the Control Office in case of abnormal movements. In case of abnormal joint movements or appearance of cracks, the system is able to detect and localize those events and automatically generate an alert. The use of distributed sensing enables sensing along the whole length of the gallery, so that no matter where the undesired events takes place, there will always be a section of sensing cable to pick up the response.

Sinkhole detection in USA

Type of Structure: Sinkhole area affecting rail and road structures.
Monitoring aim: Detect and localize impending sinkhole formations
Fiber Optic sensing technology employed: Distributed Brillouin strain sensing

Fig. 22 Distributed strain sensing cable installed by clamping in the dam inspection gallery. The sensor runs along the length of the gallery and is therefore able to pick-up deformations generated by transverse cracks crossing its path. The blue temperature sensing cable installed along the strain cable is used for temperature compensation, but can also detect water ingress in the gallery by observing rapid temperature changes

Although it is not directly a dam or levee application example, this project is included because it shows the use of distributed sensing for the detection of localized settlements that are also very relevant for levee monitoring.

A sinkhole, also known as a sink, shake hole, swallow hole, is a natural or man-made depression or hole in the Earth's surface caused by karst processes or mining activities. Sinkholes may vary in size from 1 to 600 m both in diameter and depth, and vary in form from soil-lined bowls to bedrock-edged chasms as exemplified by Fig. 23. Sinkholes may be formed gradually or suddenly, and are found worldwide. It is clear that such phenomena represent a risk for ground stability and a non-negligible

Fig. 23 Example of sinkhole formation in the area under monitoring near Hutchinson, Kansas, USA

safety risk for surface infrastructure in the surrounding areas, such as roads or rail lines. In such applications where critical area localization and the use of the discrete sensors are practically impossible because of the installation complexity and costs, a distributed sensing system is particularly suitable.

The city of Hutchinson is located in Reno County, Kansas. Hutchinson is on the route of the trans-continental, high-speed main line of one of the nation's largest railroads. The railway passes near a former salt mine well field, where mining was carried out in the early part of the twentieth century. The salt mining was performed at depths of over 400 feet by drilling wells through the shale bedrock into the thick under-ground salt beds, and then pumping fresh water into the salt, dissolving the salt to be brought back to the surface as brine, for processing and sale. This solution mining process resulted in the presence of multiple, large underground voids and caverns, which have been reported to be up to 300 feet tall and over 100 feet in diameter. In places, the shale roof rock over some of these old mine voids has collapsed, forming crater-like sinkholes that can be over 100 feet in diameter and 50 feet deep at the surface. The collapse and sinkhole formation can occur very rapidly, over a period of hours to days. Figure 23 is a photograph of a sinkhole that opened up virtually overnight at this site in 2005, by collapse of a salt cavern that was last mined in 1929. The potential rapid formation of sinkholes by collapse of old mine caverns clearly represents an issue for ground stability and a non-negligible safety risk for surface infrastructure, including the railway.

An area on the site containing old, potentially unstable salt caverns adjacent to sensitive surface infrastructure was identified with the aim of establishing an effective monitoring system in order to provide early stage detection, continuous monitoring, and automatic telemetry. Arrangements were made for alerting via cell phone and email, in case of ground deformation (strain) that may be the early signs of sinkhole formation. The distributed fiber-optic (FO) monitoring system was selected in large part because it provides thousands of monitored points using a single fiber-optic sensing cable, all measured at the same time, in a single scan (Shefchik et al. 2011). This is well-suited to defining a monitored perimeter where the exact location of where a sinkhole might form is not known precisely. In addition, this monitoring system was selected because of the ease of installation by burial in a shallow trench.

The selection of the sensing cables represents a key aspect and at the same time a challenge in the development of this project: the cable needs to be capable of withstanding hostile environmental conditions, such as wide temperature variations and burial in the earth, as well as being resistant to burrowing rodents. The cable also needs to be sensitive enough to provide early and reliable displacement detection, and capable of optimizing the transfer of forces from the ground to the fiber.

The fiber optic sensing cable is directly buried, Fig. 24, at a depth of approximately 1.4 m over a potential sinkhole area above and around salt caverns over a path with a total length of over 4 km, as shown in Fig. 25. After digging the trench, the soil is mechanically compacted, and the sensing cable deployed on the compacted soft ground before the trench is backfilled. The sensing cable is installed in several segments in order to provide easier handling during installation, and to adapt to the site by running the cable through several short, horizontally bored segments beneath

Fig. 24 Distributed strain sensor installation in a trench

a large drainage ditch, multiple road crossings, and other obstacles at the surface. All cable segments are later linked together to form a single sensing loop by FO fusion splicing; the splices between segments as well as some extra lengths of non-buried cable are stored in dedicated, above-ground junction boxes, that can be accessed for maintenance as well as for re-routing segments of cable in case a break occur due to formation of a sinkhole. An example of strain recording obtained during a sinkhole simulation is depicted in Fig. 26.

Dam core monitoring in Spain

Type of Structure: Embankment dam with clay core.
Monitoring aim: Deformation monitoring of the clay core
Fiber Optic sensing technology employed: Long-gauge interferometric sensors

The Canales Dam is built on the Genil River, in Granada, Spain. It was constructed from 1975 to 1989 and activated in 1989. The dam is 156 m tall and the crown is 340 m long. The dam is used to produce electrical power, but it is also an important tourist attraction, notably for people who like to fish. The Canales Dam controls the river flow during the year while ensuring that there is always sufficient water to keep the river "alive" even during the driest months of July and August. A view of the Canales Dam is given in Fig. 27.

The aims of the monitoring were crack detection and monitoring of the deformation of the stiff-clay core of the dam. The zone where the cracks can occur was estimated to be between 60 and 110 m deep in the clay core. This whole length was

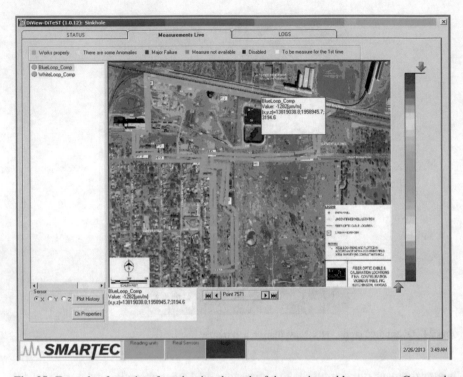

Fig. 25 Example of user interface showing the path of the sensing cable on a map. Green color indicates no strain; red color indicates 100% of the warning threshold

equipped with a 50 m long multipoint extensometer with 10 measurement zones (10 chained 5 m long deformation sensors). Preparation of the extensometer for the installation is shown in Fig. 28. The sensors were installed in a 110 m deep and 30° inclined borehole using a customized installation procedure. A mass with wheels was attached to the low-stiffness cord and slipped into the borehole. While the mass was pulling the cord into the borehole, the extensometer was attached to the cord, so the cord was used as a guide for the extensometer. The installation procedure is shown schematically in Fig. 3. After the placement of the extensometer in the desired position, the borehole was alternatively filled with grout on the anchor segments and sand was packed over the active zone. This was done to avoid perturbing the strain field in the clay core and to guarantee good deformation transfer from the clay core to each deformation sensor. Until the time of writing, no abnormal leak event has been recorded, reassuring the owner about the good performance of the watertight membrane.

Silica Alkali reaction monitoring in a concrete Dam

Type of Structure: Concrete dam
Monitoring aim: Monitor deformations induced by alkali silica reaction in concrete

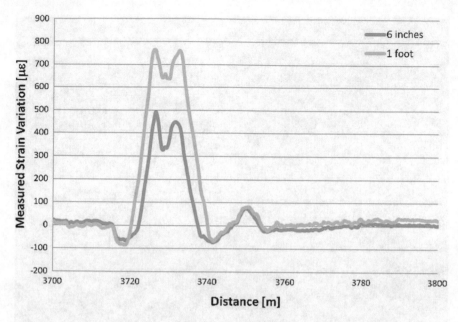

Fig. 26 Example of strain reading during sinkhole simulation test. The sensing cable is vertically moved by 6 in. and 1 foot, respectively and the resulting strain is recorded. It is possible to observe how the peak of recorded strain corresponds to the location of the test with an accuracy of a few meters. When the lateral movement is increased, more of the sensing cable records strain, indicating that a longer length of cable is being elongated by the event

Fiber Optic sensing technology employed: Long-gauge interferometric sensors

Val De La Mare reservoir (see Fig. 29) is constrained by a mass concrete dam situated on the west side of Jersey Island and was built between 1957 and 1962. The crest length is about 190 m; the dam height is 30 m above general foundation level. The dam was built in 27 monoliths (blocks) and it is owned and operated by Jersey Water. Val de la Mare dam is widely known to suffer from both Alkali Aggregate Reaction (AAR) and water seepage through existing construction joints. The dam exhibited differential displacement of some of the blocks and patches of humidity were visible on the downstream face of affected blocks. Long-gauge fiber optic deformation sensors have been installed to monitor the behavior of three blocks. The sensors have been completely integrated in the existing dam monitoring system. The fiber optic sensing system consists of 10 m long base SOFO sensors complete with a corresponding reading unit. The sensors have been installed in three of the existing vertical drain holes in blocks 11, 15 and 18. The location of the drain holes is shown in red on the right of Fig. 30. The sensors are 10 m in length and two sensors were joined mechanically together, effectively creating one 20 m sensor, placed in an 80 mm diameter drain hole. The output from reading unit has been linked to the Jersey Water SCADA system to provide a visual output showing historic trends. Multi-year data acquisition and multi-parameter regression analysis is used to evaluate the effects

Fig. 27 Canales Dam, Spain

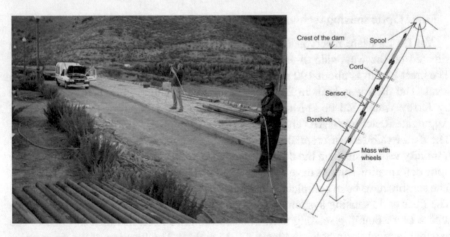

Fig. 28 Installation of long-gauge strain sensors in a borehole

Fig. 29 Val De La Mare dam, Jersey Island

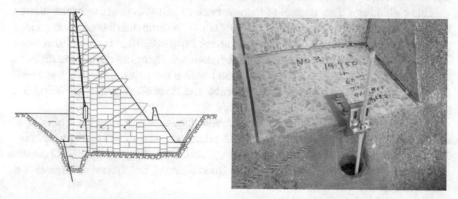

Fig. 30 Location of long-base extensometers in the inclined boreholes between the dam crest and the inspection gallery (red lines). Picture shows a detail of the upper anchor point installation

of alkali-silica reaction, aiming to separate its effects from the one of cyclic actions such as seasonal temperature and water-level variations.

Application to Tailing Dams

El Mauro Tailing dam in Chile

Type of Structure: Mining tailings dam
Monitoring aim: Monitor long term deformations and pore water pressure

Fiber Optic sensing technology employed: Long-gauge interferometric sensors and Fabry-Perot piezometers

The operation of tailing dams, typical of mining operations, in their construction phases includes the hydraulic deposit of the sand, which will form the retaining structure of the reservoir. This deposit phase is programmed, forming thin layers of clean sand which is further compacted. This volume dries out the excess water to accept the next layer. This system has shown to be cost- and quality-effective, and has been used for the last century, changing from an upstream to a downstream deposition method, which has proven to be safer for the retaining structure under seismic conditions. The area of deposit of the sand is moved along the dam in order to assess the dissipation of the transport water, which has to evaporate and/or drain into the underlying compacted sand strata. At the initial state of the dams, the sand deposition is not simple for the operators, as the cyclone plant, which selects the coarse sand fraction from the fine silty slime, has an excess elevation pressure due to their physical locations, the deposit area is very small, the runs for dissipating the excess water are very short and the surplus water tends to accumulate over the horizontal drains, with possible clogging of the open drainage matrix.

This called for a tight control of the presence of water in the base of the dam by means of 10 fiber optic piezometers and 2300 m of DiTemp distributed armored temperature cable, located some 2 m above drainage layers (Fahrenkrog and Fahrenkrog 2012). The hydraulic conditions of a tailing dam are changing during the different construction stages and have to be addressed with a monitoring layout, which will allow the most of the variables to be recorded and controlled during the active and passive phase of the basin.

In order to control the stability and prevent landslides, 9 long base fiber optic sensors have been installed in the body of the dam. The sensor has been inserted in a special structure that adheres completely to the soil. Three long-gauge SOFO sensors are joined to form a tridimensional gauge for settlement and lateral deformation as shown in Fig. 31.

Fig. 31 Installation of long-gauge extensometers in the body of the tailing dam. Courtesy of Geomediciones

Additionally, fiber optic piezometers have been installed at several locations in the dam body. Those sensors were selected because of their insensitivity to damage produced by lightning strikes that are common in the area. Additionally, optical fiber sensors allow the use of cables longer than 1 km, without loss of performance. This enables installation in very large structures such as those tailing dams. Optical connection cables can also be lighter and less expensive than the equivalent copper wires used to connect conventional sensors over long distances.

Conclusions

The monitoring of new and existing structures is one of the essential tools for modern and efficient management of the infrastructure network. Sensors are the first building block in the monitoring chain and are responsible for the accuracy and reliability of the data. Progress in sensing technologies comes from more accurate and reliable measurements, but also from systems that are easier to install, use and maintain. In recent years, fiber optic sensors have taken the first steps in structural monitoring, particularly in civil and geotechnical engineering. Different sensing technologies have emerged and evolved into commercial products that have been successfully used to monitor hundreds of structures. No longer a scientific curiosity, fiber optic sensors are now employed in many applications where conventional sensors cannot be used reliably or where they present application difficulties.

The use of distributed fiber optic sensors for the monitoring of civil and geotechnical structures opens new possibilities that have no equivalent in conventional sensors systems. Thanks to the use of a single optical fiber with a length of tens of kilometers it becomes possible to obtain dense information on the strain and temperature distribution in the structure. This technology is therefore particularly suitable for applications at large or elongated structures, such as dams, dikes, levees, large bridges and pipelines.

Three characteristics of fiber optic sensors can be highlighted as the reasons of their present and future success: (1) the precision of the measurements; (2) the long-term stability and durability of the fibers; and (3) the possibility of performing distributed and remote measurements over distances of tens of kilometers.

References

Sills, G.L., Vroman, N.D., Wahl, R.E. and Schwanz, N.T., 2008. Overview of New Orleans levee failures: lessons learned and their impact on national levee design and assessment. Journal of Geotechnical and Geoenvironmental Engineering, 134(5), pp. 556–565.
Inaudi, D. and Glisic, B., 2006, July. Distributed fiber optic strain and temperature sensing for structural health monitoring. In Proceedings of the Third International Conference on Bridge Maintenance, Safety and Management, Porto, Portugal (pp. 16–19).

Udd, E. and Spillman Jr, W.B. eds., 2011. Fiber optic sensors: an introduction for engineers and scientists. John Wiley & Sons.

Glisic, B. and Inaudi, D., 2008. Fibre optic methods for structural health monitoring. John Wiley & Sons.

Inaudi, D. and Glisic, B., 2006, March. Reliability and field testing of distributed strain and temperature sensors. In Smart Structures and Materials 2006: Smart Sensor Monitoring Systems and Applications (Vol. 6167, p. 61671D). International Society for Optics and Photonics.

Inaudi, D., 2004, June. Testing performance and reliability of fiber optic sensing system for long-term monitoring. In Second European Workshop on Optical Fibre Sensors (Vol. 5502, pp. 552–556). International Society for Optics and Photonics.

Pinet, É., 2009. Fabry-Pérot fiber-optic sensors for physical parameters measurement in challenging conditions. Journal of sensors, 2009.

Rodrigues, C., Inaudi, D., Juneau, F. and Pinet, É., 2010. Miniature Fiber-Optic MOMS Piezometer. Geotechnical News, 28(3), p. 24.

Inaudi, D., Elamari, A., Pflug, L., Gisin, N., Breguet, J. and Vurpillot, S., 1994. Low-coherence deformation sensors for the monitoring of civil-engineering structures. Sensors and Actuators A: physical, 44(2), pp. 125–130.

Glišić, B., Inaudi, D., Lau, J.M., Mok, Y.C. and Ng, C.T., 2005, December. Long-term monitoring of high-rise buildings using long-gage fiber optic sensors. In Proceedings of 7th International Conference on Multi-Purpose High-Rise Towers and Tall Buildings, Dubai, UAE (pp. 10–11).

Glisic, B., Inaudi, D. and Nan, C., 2002. Pile monitoring with fiber optic sensors during axial compression, pullout, and flexure tests. Transportation Research Record: Journal of the Transportation Research Board, (1808), pp. 11–20.

Inaudi, D. and Casanova, N., 2000, June. Geostructural monitoring with long-gage interferometric sensors. In Nondestructive Evaluation of Highways, Utilities, and Pipelines IV (Vol. 3995, pp. 164–175). International Society for Optics and Photonics.

Dunnicliff, J., 1993. Geotechnical instrumentation for monitoring field performance. John Wiley & Sons.

Inaudi, D., & Glisic, B. (2005, May). Development of distributed strain and temperature sensing cables. In 17th International Conference on Optical Fibre Sensors (Vol. 5855, pp. 222–226). International Society for Optics and Photonics.

Inaudi, D., & Church, J. (2011). Paradigm shifts in monitoring levees and earthen dams distributed fiber optic monitoring systems. In 31st USSD Annual Meeting & Conference, San Diego, California, USA.

Inaudi, D., Casanova, N., Martinola, G., Vurpillot, S., & Kronenberg, P. (1998, October). SOFO: Monitoring of Concrete Structures with Fiber Optic Sensors. In 5th International Workshop on Material Properties and Design, Weimar (pp. 495–514).

Thévenaz, L. (1999, September). Monitoring of large structure using distributed Brillouin fibre sensing. In 13th International Conference on Optical Fiber Sensors (Vol. 3746, p. 374642). International Society for Optics and Photonics.

Wang, M. L., Lynch, J. P., & Sohn, H. (Eds.). (2014). Sensor Technologies for Civil Infrastructures, Volume 2: Applications in Structural Health Monitoring. Elsevier.

Artières O., Beck Y.L., Khan A.A., Cunat P., Fry J.J., Courivaud J.R., Guidoux C., Pinettes P. (2010). Assessment of Dam and Dikes behaviour with a fiber optic based monitoring solution. Proc. of the 2nd Dam Maintenance Conference, Zaragoza, November 2010, pp. 79–86.

Shefchik, B., Tomes, R., & Belli, R. (2011). Salt Cavern Monitoring System for Early Warning of Sinkhole Formation. Geotechnical News, 29(4), 30.

Fahrenkrog, C., & Fahrenkrog, A. (2012). Instrumentation with fiber optic sensors emphasizing operation of tailing dams and landfill monitoring. In 5th European Conference on Structural Control, Genoa, Italy.

Application of the Helicopter Frequency Domain Electromagnetic Method for Levee Characterization

Adam Smiarowski, Greg Hodges and Joe Dunbar

Abstract Levee characterization requires many miles of ground to be surveyed. Remote sensing methods, particularly those capable of installation on an aircraft, are capable of quickly surveying large areas. The helicopter frequency domain electromagnetic technique (HEM) involves towing an electromagnetic transmitter and receiver that measure signals proportional to the electrical conductivity of the ground. Information about the electrical conductivity of the ground can be used to make inferences about the distribution of soils or rocks in the subsurface. HEM data can be interpreted by correlating the apparent conductivity to soil or rock type, as well as for looking for lateral and depth extent of anomalous zones. HEM data provide depth information by performing measurements at different frequencies. Here, we provide a brief review of the frequency domain method, highlighting the transformation of HEM data to apparent resistivity, which is critical for interpretation. We then discuss two case histories using an HEM system for levee characterization and hazard detection. With clay being the primary building material for the levees, the minimum layer thickness that can be accurately resolved with an upper frequency of 140 kHz is about 1 m. The HEM data are particularly useful at detecting anomalous zones for follow up investigation. These zones were caused by underground channels (which provide pathways for water flow) and in one case by significant cracking of a levee. The examples provided here highlight the utility of HEM surveying for levee characterization.

A. Smiarowski (✉)
CGG, Toronto, Canada
e-mail: adam.smiarowski@cgg.com

G. Hodges
Sander Geophysics, Ottawa, Canada
e-mail: ghodges@sgl.com

J. Dunbar
Engineer Research and Development Center, Vicksburg, USA
e-mail: Joseph.B.Dunbar@erdc.dren.mil

© Springer Nature Switzerland AG 2019
J. Lorenzo and W. Doll (eds.), *Levees and Dams*,
https://doi.org/10.1007/978-3-030-27367-5_6

Introduction

Routine application of the active source airborne electromagnetic technique dates back to the 1950s (Palacky 1981). The first systems were used primarily for hard rock mineral exploration and operated in the frequency domain, at a single frequency, with the main goal being "bump detection", that is, looking for isolated anomalies in the recorded signal with the hope of discovering an isolated target (Fountain 1998). As knowledge and experience increased, and technology improved, the frequency domain technique has become much more sophisticated in terms of application, processing and goals. Current systems operate at multiple frequencies (with each frequency providing information about a different depth), are calibrated and used to detect subtle features of interest to high-resolution engineering and geotechnical applications.

In the frequency domain electromagnetic (FDEM) method, an electromagnetic transmitter generates a sinusoidal EM field at discrete frequencies. An EM receiver measures the electromagnetic response from the ground, which can then be processed and interpreted to yield information about the distribution of electrically conductive material in the earth's subsurface. The helicopter FDEM technique has sensitivity from surface to a maximum of about 150 m, depending on the frequency range applied and ground conductivity. The frequency-domain electromagnetic method has application in geological mapping, mineral prospecting, hydrogeology, permafrost mapping, slope stability, environmental contaminant mapping, unexploded ordnance detection, engineering infrastructure planning, as well as others. Numerous authors have discussed the theory of the method (Grant and West 1984; West and Macnae 1991), providing derivations for equations to calculate the response of a layered earth, as well the resolution of HEM and its assistance with various problems (groundwater-Siemon 2006; sea-ice thickness—Reid et al. 2003). Here we briefly review the physics describing the method and describe standard processing techniques which transform from the "data space" to the "model space", which facilitates interpretation of the data.

The scope of this paper is to describe the essential theory and application of HEM system to levee characterization. The physical property of interest for HEM systems is electrical conductivity (and its reciprocal, resistivity); we describe how conductivity measurements can be used to provide information about levee stability. We then describe the RESOLVE FDEM instrument and its capabilities. The next section describes HEM data processing and the steps involved in transforming field data to conductivity estimates. Interpretation of conductivity is the essential link to determining levee stability; we conclude this paper with a number of case studies highlighting the utility of HEM for levee characterization.

Resistivity of Common Materials

Electrical conductivity σ [S/m] describes the ability of a material to conduct electricity. Electrical resistivity ρ [Ω m] = 1/σ, the reciprocal of conductivity, describes how strongly a material opposes an electric current. Electrical resistivity values of earth materials vary over more than seven orders of magnitude (Palacky 1988). Resistivity is affected by the matrix conductivity of the material, as well as strongly dependent on pore volume, pore saturation and pore fluid conductivity. The variation in material conductivity makes mapping from resistivity measurements to geologic material difficult. Rocks can have variable mineral content; soils can be classified as "sandy-clay" or "clayey-sand" because of differing composition; rocks can be weathered, fresh, have variable porosity, saturation, compaction (influencing degree of electrical connectivity) etc. Figure 1 shows the resistivity of some common earth materials. For levee investigations, soils such as clay, till, sand and gravel are of particular interest. Note that clay is generally more conductive than tills and sands. Any a priori knowledge is very useful in helping to constrain interpretation from resistivity to soil type. Borehole logs are generally very detailed, differentiating between clayey-sand and sandy-clay at fine intervals. Because of overlapping resistivity ranges, it can be very difficult to uniquely identify such materials by electrical conductivity alone.

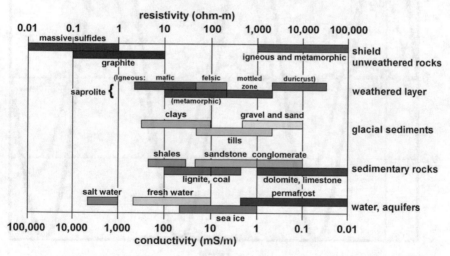

Fig. 1 Resistivity and conductivity of some common geologic materials. From Palacky (1988)

The Frequency Domain Electromagnetic Method

In the FDEM method, a transmitter coil generates a time-varying sinusoidal EM signal at a particular frequency. According to Faraday's law of induction, a time-varying magnetic field induces a voltage and causes a current to flow in the earth. Ohmic losses in the ground cause the impressed current (and the corresponding magnetic field picked up at the receiver coil) to lag behind the transmitter signal. The signal at the receiver is shown in Fig. 2. The transmitter signal is used as a timing reference for the measured field. In practice, the primary signal at the receiver is electromagnetically cancelled by a bucking coil which results in the receiver measuring the secondary signal from the ground. The secondary signal is decomposed into components in-phase and out-of-phase (quadrature) with the reference. The ability to measure the in-phase signal is one distinguishing feature between frequency-domain and time-domain airborne electromagnetics.

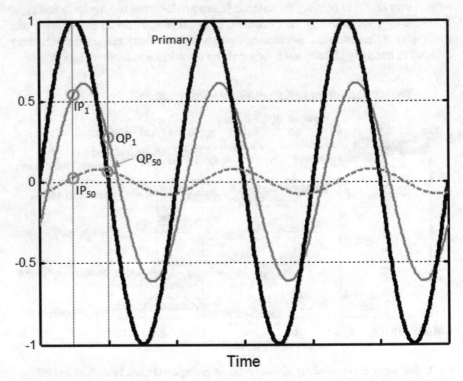

Fig. 2 Signals measured by the receiver. The black line shows the primary signal (which is used as a timing reference); the gray line shows the signal from the ground (secondary) for a 1 Ω m halfspace (solid) and 50 Ω m halfspace (dashed). The secondary signal is very small compared to the primary and has been scaled up by 250 times for viewing. The processing streams extracts components in-phase (IP) and out-of-phase (quadrature-QP) with the reference signal

The frequency domain response to a layered earth is provided by Frishknecht (1967), Ward (1967) and Ward and Hohmann (1988). For a horizontal coplanar coil set at height h above a horizontally layered ground, the secondary magnetic field H_s is given by

$$H_s = H_0 \int\limits_0^\infty R(\lambda)\lambda^2 \exp(-2u_0 h) J_0(\lambda s) d\lambda \tag{1}$$

where H_0 is the primary magnetic field (which can be obtained with knowledge of the transmitter-receiver separation and orientation), s is the transmitter-receiver coil separation, λ is the integration variable, and J_0 is the zero-order Bessel function of the first kind. The reflection term $R(\lambda)$ can be written as

$$R(\lambda) = \frac{Y_1 - Y_0}{Y_1 + Y_0} \tag{2}$$

where $Y_0 = u_0/i\omega\mu_0$ is the intrinsic admittance of free space (air, approximately) and Y_1 is admittance at the surface of the earth, i is the imaginary number, ω is angular frequency, μ_0 is magnetic permeability of free space and $u_0 = \left(\lambda^2 + \kappa_0^2\right)^{1/2}$, where k_0 is the wave number of free space. For a layered earth with N layers, the surface admittance Y_1 can be obtained as:

$$Y_l = \widehat{Y}_l \frac{Y_{l+1} + \widehat{Y}_l \tanh(u_l t_l)}{Y_l + \widehat{Y}_{l+1}\tanh(u_l t_l)}, \quad l = 1, 2, \ldots L - 1 \tag{3}$$

where $\widehat{Y}_n = \frac{u_n}{i\omega\mu_0 u_n}$ and $u_n = (\lambda^2 + k_n^2)^{1/2}$
t_n is the thickness of the nth layer, and σ_n is the layer conductivity.

The wavenumber k_n, invoking the quasi-static limit and assuming magnetic permeability of free space, is given by

$$k_n = (i\omega\sigma_n\mu_0)^{1/2}$$

In the bottom layer (layer $n = N$), $Y_N = \widehat{Y}_N$. Using this relation, Eq. (3) can be solved recursively to obtain Y_1 and the reflection coefficient R in Eq. 2. This series of equations can be used to calculate the in-phase and quadrature HEM response to a layered earth.

The in-phase and quadrature information can be transformed to an amplitude/phase relationship as (Fraser Fraser and Geophysics 1978):

$$\|\Lambda\| - \sqrt{IP^2 + QP^2} \quad \emptyset = \tan^{-1}\frac{QP}{IP} \tag{4}$$

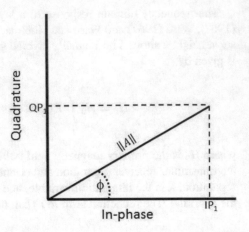

Fig. 3 A pictorial depiction of the relationship between in-phase (IP) and quadrature (QP) components and the amplitude ‖A‖ and phase (ϕ)

The ratio of the in-phase to quadrature signal is the phase angle Ø and is representative of the conductivity of the ground; as conductivity of the ground increases, response is more in-phase and the calculated phase angle decreases. The relationship between in-phase and quadrature with amplitude and phase is shown in Fig. 3.

The amplitude $\|A\|$ is very sensitive to the height of the system above the ground (Fraser 1978), with an approximately $1/h^2$ dependence on height. A small error in measured system altitude can cause a large change in signal levels. Figure 4 shows amplitude and phase as a function of system altitude. At 35 m altitude an amplitude measurement of 1650 pm is obtained for a halfspace of 85 Ω m; a 3 m altitude error (to a reading of 32 m) caused by, say, vegetation would result in a calculated apparent resistivity of 170 Ω m. Using phase angle to calculate apparent resistivity, the same altitude error would give an apparent resistivity of 72 Ω m. Phase angle provides apparent resistivity calculations with more tolerance to altitude errors than amplitude-derived apparent resistivity, as would be done in time-domain electromagnetics.

The Resolve FDEM System

Figure 5 shows a picture and schematic diagram of the RESOLVE® frequency domain electromagnetic system as currently implemented. RESOLVE has 6 transmitter coils (horizontal co-planar coils, most sensitive to flat-lying features, range in frequency from 400 to 140,000 Hz. There is 1 vertical co-axial coil with best sensitivity to vertical features operated at 3300 Hz).

A standard processing step for the RESOLVE system is to calculate the apparent resistivity from the measured in-phase and quadrature data. Besides the obvious value of a rock property for interpretation, this step is important for ensuring correct calibration and levelling of the data. For the measured survey altitude, a homogenous halfspace model is assumed and a look-up table of in-phase and quadrature signal (or phase and amplitude) is computed using Eq. (1) for each frequency at a range of

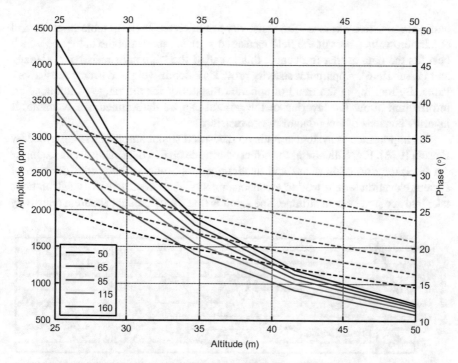

Fig. 4 Comparison of sensitivity of amplitude (left axis; solid lines) and phase (right axis; dashed lines) to altitude for various homogenous halfspace resistivity values calculated for a 140,000 Hz coplanar transmitter-receiver separated by 7.9 m

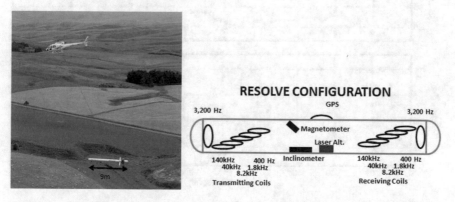

Fig. 5 A picture of the RESOLVE system in flight (left) and a schematic diagram of the RESOLVE frequency domain EM system. The 9 m long "bird" houses transmitter and receiver coils and is towed by the helicopter at a nominal altitude of 30 m above the ground (or obstacles such as trees). Usually a magnetometer is also carried on the bird, as well as a laser altimeter to provide height above ground and an inclinometer to correct for changes in coil altitude

halfspace resistivity values. The resistivity at which the look-up table in-phase and quadrature values best fit the field measured signals represents the halfspace which best fits the data at that frequency; this is called the "apparent resistivity" (Huang and Fraser 1996). Apparent resistivity provides a depth-weighted average resistivity value. By looking at the trend of apparent resistivity for the range of frequencies, information about the layering of the ground can be determined, for example, if layering consists of a conductor over a resistor.

The importance of system altitude on measured signal amplitude is described by Fraser (1978); Fig. 6 illustrates the effect of altitude on EM data using a data example. An interpretation of in-phase and quadrature amplitude without consideration of system elevation would lead to erroneous results. Apparent resistivity calculations take into account system altitude. The result is that the calculated apparent resistivity

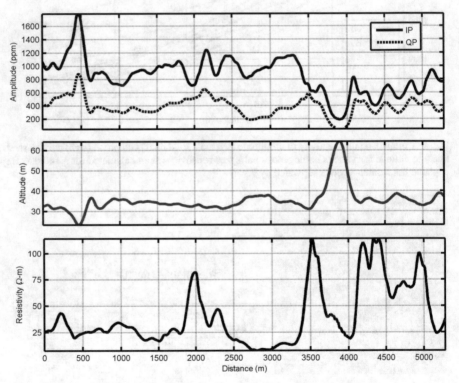

Fig. 6 The top panel shows measured in-phase (solid) and quadrature (dashed) data while the middle panel shows system altitude. The bottom panel shows calculated apparent resistivity. At position 500 m, there is a sharp increase in measured data; at position 3900 m, there is a broad decrease in signal. All else equal, one would conclude increased conductivity at 500 m and increased resistivity at 3900 m; however, in this case, the measured amplitude is an altitude effect. The middle panel of Fig. 6 shows the system altitude; note that at 500 m, there is a sharp decrease in altitude. The altitude decrease means the system is closer to the current distribution in the ground, which causes an increase in EM amplitude. The situation is opposite for position 3900 m; the altitude increases, decreasing the measured signal

is about the same at 500 as 3900 m, even though the EM amplitude is quite different. Figure 6 shows why interpretation of profile EM data can be misleading and why conversion to resistivity is critical.

Depth of Investigation

A rough estimate of the depth of investigation is provided by using the electromagnetic skin depth, where the EM field decays to a value 1/e at the skin depth δ (Spies 1989) of

$$\delta \approx 503\sqrt{\frac{\rho}{f}} \tag{5}$$

where ρ is the material resistivity and f is the frequency of the EM field. When determining the depth of investigation, experience has shown that the RESOLVE frequency-domain system operating at its nominal survey height of 30 m has maximum sensitivity at about 0.75 δ. Using inversion of all frequencies, depth of investigation has extended to as much as 2δ where uniform upper layers are present. Approximate depth of exploration for the frequencies of the RESOLVE system is shown in Fig. 7; for levees composed of clays, sands and soils, resistivity is expected to range from 1 to 200 Ω m (as in Fig. 1). In some surveying applications, we are

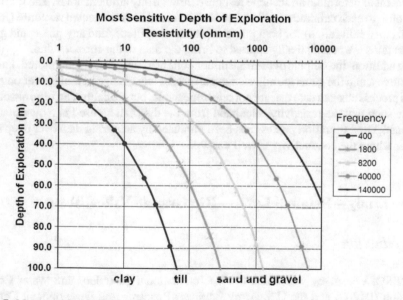

Fig. 7 Most sensitive depth of exploration, calculated using 0.5 skin depths. The typical resistivity of material of interest for the levee case is indicated at the bottom of the figure

interested in detecting thin layers. When performing an advanced imaging or inversion of the data, the minimum layer thickness that can be detected under optimal conditions is about 0.2 δ. This means that the minimum conductive layer that can be detected is about 1 m and minimum till layer about 2.5 m. Note also that the low to mid frequencies are sensitive to layers thicker than a typical levee, so they will not provide information about the levee itself but material beneath it.

System Calibration for of the RESOLVE System

Modern instrumentation and processing techniques have allowed the FDEM method to acquire calibrated, repeatable data that can be used in an absolute sense to provide information about rock and soil properties. Calibration is an essential data acquisition step to provide a base-level for the data.

Calibration of the system occurs during the survey, using an automatic, internal calibration process. At the beginning and end of each flight, and at intervals during the flight, the system is flown up to high altitude to remove it from any "ground effect" (ensuring the response from the earth is zero). Any remaining signal from the receiver coils is measured as the zero level, and is removed from the data collected until the time of the next calibration. Following the zero level setting, internal calibration coils, for which the response phase and amplitude have been determined at the factory, are automatically triggered—one for each frequency. The on-time of the coils is sufficient to determine an accurate response through any ambient noise. The receiver response to each calibration coil "event" is compared to the expected response (from the factory calibration) for both phase angle and amplitude, and any phase and gain corrections are automatically applied to bring the data to the correct value.

In addition, the outputs of the transmitter coils are continuously monitored during the survey, and the gains are adjusted to correct for any change in transmitter output. This process ensures data that are not only repeatable, but are calibrated in an absolute sense such that the resistivity calculated from the data can be used in a quantitative manner. The calibration process has been validated by acquiring data over deep salt water where the conductivity is well known.

Case Study—Retamal Levee—Rio Grande Valley, Texas

Introduction

A RESOLVE survey was flown for the International Boundary and Water Commission (IBWC) and the U.S. Army Engineer Research and Development Center (ERDC) along the lower reaches of the Retamal Levee, Rio Grande Valley (RGV), Texas. The survey goal was to assist in monitoring of the flood control levees and to provide early identification of potential hazards. A traditional engineering assessment using evenly spaced bores was considered but would have taken years to complete. HEM was used to gain a better understanding of the levee system in a short amount of time. Three HEM lines were flown; one over the top of the levee, and one each 50 m inside and outside the levee. The operating frequency ranged from 380 to 102,000 Hz.

Discussion

The HEM data are used to calculate the apparent resistivity (or its inverse, apparent conductivity) of the ground at each frequency, which are then compiled into maps. The high-frequency apparent conductivity for the survey area is shown in Fig. 8; inset images provide examples of detail and conductivity contrasts. Apparent conductivity was used to generate a map of interpreted soil material, with conductive areas mapped as clay and resistive areas mapped as sands. The levee is generally thin and narrow compared to the skin depth of the transmitted frequencies and the HEM is primarily sensitive to the shallow layers beneath the levee. Conductivity anomalies were identified as "continuous", "isolated" or "cross-cutting" to assess the risk to the levee system. For the most part, HEM surveys in the RGV were successful

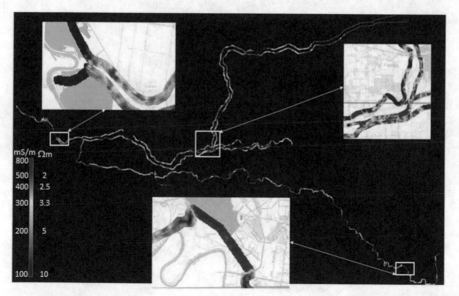

approximately 75 miles

Fig. 8 Calculated apparent conductivity/resistivity for the RESOLVE system in the survey area. The inset images are included to show a few detail images and highlight that strong conductivity/resistivity variations occurred along the levees

in accurately determining foundation soils beneath the levee as evidenced by soil borings and cone-penetrometer tests (CPTs) data. Sorensen and Chowdhury (2010) found that geophysical data (both DC resistivity and HEM data) could not identify levee composition, but both were useful in evaluating the continuity of deposits and to assess the extent of anomalous foundation conditions. HEM resistivity profiles indicated similar patterns to borehole drilling and were used to help estimate the extent of channel deposits and to evaluate appropriate improvement measures.

Figure 9 shows the calculated apparent conductivity from RESOLVE's high frequency in the vicinity of the Retamal Levee. The Retamal Levee section was constructed using clay materials, but the RESOLVE survey showed an area to have an anomalously low conductivity/high resistivity (Dunbar et al. 2005). A site inspection revealed that the levee was cracked and desiccated (Dunbar et al. 2005). The RESOLVE survey, flown in 2001, was conducted at the end of a 4-year period of low rainfall (Dunbar et al. 2005). Rainfall is shown in Fig. 10. Insufficient rainfall can lead to the levee construction material (mainly clay) becoming quite dry and showing significant cracking, which can be "healed" with rain fall and re-moisturization. The high resistivity mapped by the HEM is likely due to the very dry conditions. This is one reason why it can be difficult to directly map soil type from HEM data. The electrical connectivity of the material can change with dry or wet conditions, which affects the bulk electrical conductivity and the EM measurements. Assuming

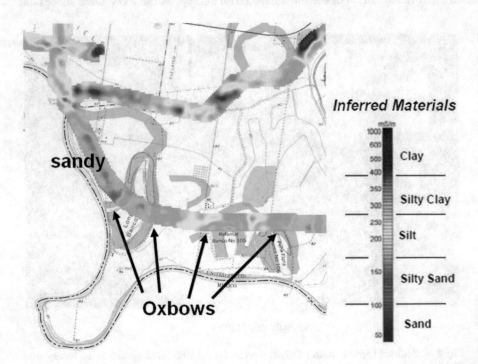

Fig. 9 Apparent conductivity calculated from RESOLVE high frequency data in the vicinity of the Retamal levee. The southern area was unusually resistive considering the building material was primarily clay. Known oxbow lakes are highlighted in light blue

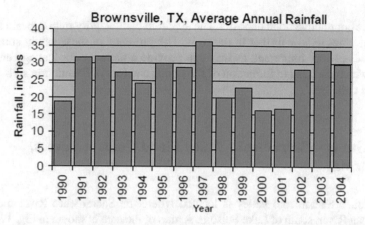

Fig. 10 Annual rainfall in inches for Brownsville Texas (situated at south-east corner in Fig. 7), between 1990 and 2004 (reproduced from Dunbar, Report 5). Data from NOAA, NWS 2005

no change in levee composition, HEM surveys could be used to detect resistivity changes that can occur due to moisture levels, which may be useful for assessing changes to levee health over time.

Figure 11 shows a detailed section of the Rio Grande Levees survey. The higher resistivity (blue) river sand deposits inside the oxbow are quite obvious, relative to

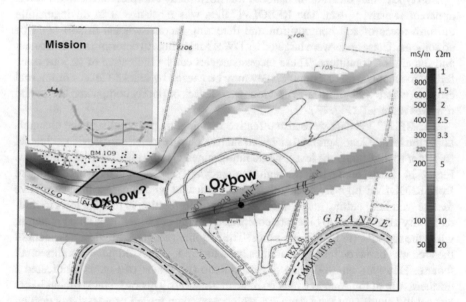

Fig. 11 A detailed section of the Rio Grande Levees survey. The levees are shown by the thin lines inside the colored apparent conductivity. The resistive (blue) material inside the oxbow has been interpreted to be transported sands, while the more conductive areas are flood deposits. The oxbow to the left in the northern flight lines was not known to exist before the HEM survey

the more conductive flood deposits. The apparent conductivity map appears to show another, older oxbow further to the west. The presence of these sandy soils under the levees is very important, as they can provide a pathway for leakage under the levee, which can lead to formation of sand boils and loss of foundation soils leading to levee collapse.

Case Study—Sacramento Valley Flood Control Levees, California

California's Sacramento valley is located where the Sacramento River meets the American River, south of Lake Fulsom. A map of the area is shown in Fig. 12. These water bodies make Sacramento susceptible to flooding. As part of the long term goal of providing flood protection, the Department of Water Resources, California evaluated 350 miles (560 km) of levees. Information about potential problems with the levee was desired. The RESOLVE frequency domain system was used to assist in this evaluation, with three parallel survey lines flown along a portion of the levee system, collecting a total of 453 line-miles of data. The system was flown at an altitude of 30 m, operating with 5 coplanar frequencies spaced logarithmically between 400 Hz and 140 kHz.

Soil type of the construction material was not directly interpreted from RESOLVE apparent resistivity data. The RESOLVE data was most useful in distinguishing uniform areas of soil composition and detecting inconsistencies in soil or levee conditions. These areas were targeted by DWR for borings to determine embankment and foundation conditions. These inconsistencies can be indicative of various levee failure mechanisms. For example, they may be caused by cracked levees or internal erosion (where water infiltrates pervious sand levees or poorly compacted levees and removes material) (Fig. 13).

The RESOLVE data were also effective in detecting potential seepage channels under the levee. Coarse grained material, such as sand deposits, can provide a channel for water to flow under the levee, while fine-grained clays usually restrict water flow. Figure 14 shows, on the left, the high frequency apparent conductivity as calculated from RESOLVE high frequency data. The indicated area shows a linear resistive feature (red line) cutting across the levee. The interpretation is that the low- to mid-resistivity areas (green to blue) are comprised of clay or other fine-grained material, while the resistive (red) areas are sands. The original geomorphologic assessment of the area was based on historical studies back to 1935, which did not show this small feature. However, an 1892 map of Sacramento (right side of Fig. 14) indicated a northern arm of the Arcade Creek (Amine et al. 2009). This old river segment likely left behind significant sand deposits, which were then buried progressively deeper, and now form a potential seepage path and comprise a risk to levee stability.

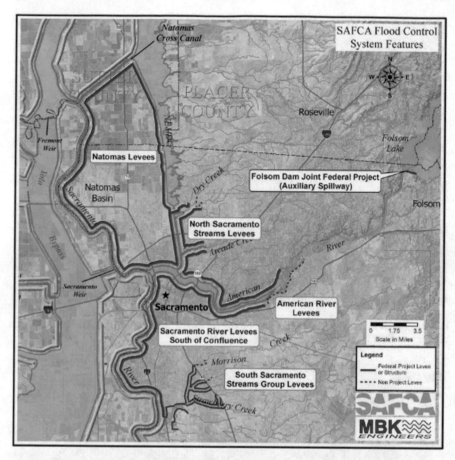

Fig. 12 Map of Sacramento, California, with levees and nearby water bodies. The Sacramento River runs north-south and meets the east-west American River just north of Sacramento. From Sacramento Area Flood Control Agency (2018)

As shown in Fig. 15 (a 3D rendering of the resistivity shown in Fig. 14), there is a thin resistive band (interpreted to be a sand deposit) cutting across the levee. This is presumably the old river tributary described previously. As well, a thick deposit of resistive material has accumulated just down-stream of where the Arcade Creek meets the river; this is interpreted to be a sand deposit. The information from the RESOLVE survey led to the length of a proposed remediation cut-off wall being extended to this area in order to mitigate possible under-seepage (Amine et al. 2009).

When performing seepage analysis, a landside blanket of impermeable clay significantly influences seepage exit gradients and slope stability factor. A waterside blanket will also affect these factors. While the landside blanket can be assessed through boring or ground-based geophysics, performing the same measurements on the waterside can be difficult. If the waterside blanket cannot be verified, protocol

Fig. 13 Aerial view over a portion of the survey area near Sacramento. The red lines show the flight lines of the HEM system (and location of the levees)

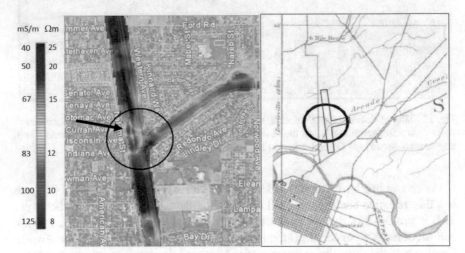

Fig. 14 High-frequency apparent conductivity (red is resistive, blue is conductive) over the Sacramento levee shown in the left panel. The resistivity anomaly (indicated by black arrow) from the airborne EM data suggested an old river channel which was not shown on current maps. This prompted a search into the available records and a 19th century map confirmed the location of a small tributary of Arcade Creek (shown on the right panel; red lines show survey area) Modified from Amine et al. (2009)

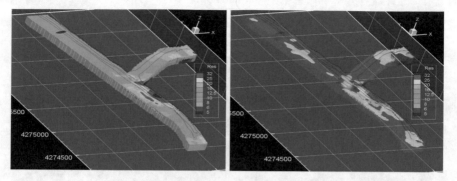

Fig. 15 shows in the left panel a rendering of the RESOLVE resistivity data at Arcade Creek (Amine et al. 2009). The right side shows the same image after stripping away the conductive material, leaving the relatively-resistive sand channels. The northern section of the creek likely has thinner sand channels than the southern part

is to assume the blanket is absent when performing calculations, which decreases the safety factor calculated and may not be an accurate representation of the levee (Amine et al. 2009). The HEM data can be used to determine if the waterside blanket is present.

Figure 16 shows an example of the vertical resistivity section calculated for the waterside of the levee for two different areas. In the top figure, the HEM data show a resistive layer (25 Ω m) over a conductive layer (5 Ω m). Using the resistivity intervals of Fig. 1 this is interpreted to be till/sand over clay. Borehole drilling indicates only very thin at-surface fine-particle material, then a thick sequence of coarse (clayey-sand) material; there is no blanket layer present. In the bottom panel, the sand layer has a thick (4–6 m) blanket layer of clay overtop, corroborated by the borehole drilling, which shows a thick sequence of fine-grained material and then coarse-grained material at depth. Incorporating these results will provide a more accurate indication of levee performance (Amine et al. 2009).

Conclusions

Airborne frequency domain methods (HEM) are capable of obtaining electrical conductivity information about the earth from about the top 1 to 100 m below surface. Data are typically transformed to apparent conductivity, which removes variations in system altitude and allows easier interpretation of ground material. For levee characterization the HEM-derived conductivities, mapped in 3D, give indications of the locality and extent of sand channels and clay layers, and may give an idea of the levee construction material. Follow-up boring plans can be optimized based on the HEM information to be more cost-effective, and the results used to refine the interpretation. In the case studies shown, HEM data were effective at detecting sandy

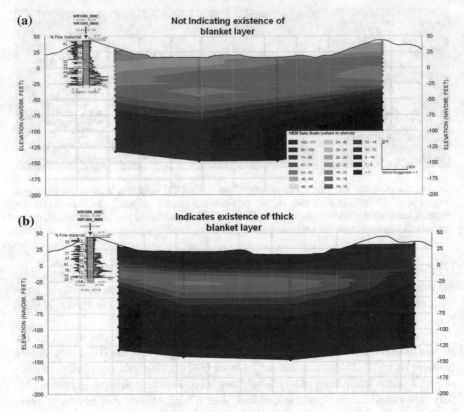

Fig. 16 Apparent resistivity calculated from HEM data and plotted as a vertical resistivity depth section. Top panel **a** shows an area where no clay blanket is present; the area is sand over clay. Bottom panel **b** shows where a blanket layer of clay is present on the waterside of the levee. Inset figure shows interpreted soil material from borehole drilling (CL—clay; SC—clayey-sand; ML—silt) along with percentage of fine material. From Amine et al. (2009)

channels and determining their spatial extent, including old oxbows and buried river channels that provide seepage pathways under the levee, risking sand boils or levee collapse from foundation erosion. In another example high resistivity values from the HEM data indicated dry, sandy conditions, and led to discovery of significant cracking in the levee due to desiccation of the levee material. HEM surveying provides a rapid method in which to assess levee hazards and provide information for levee characterization.

References

Amine, D., Hodges, G. H., Selvamohan, S., and Marlow, D. Correlating Helicopter EM and Borings for Levee Evaluation Studies in California. Symposium on the Application of Geophysics to Engineering and Environmental Problems 2009: pp. 126–134.

Dunbar, J.B., Llopis, J.L., Sills, G.L., Smith, E.W., Miller, R.D, Ivanov, J., and Corwing, R.F. Condition Assessment of Levees, U.S. Section of the International Boundary and Water Commission. Report 5 Flood Simulation Study of Retamal Levee, Lower Rio Grande Valley, Texas, Using Seismic and Electrical Geophysical Models. Army Corps of Engineers, 2005.

Fraser, D. C., Resistivity mapping with an airborne multicoil electromagnetic system, Geophysics, 43, 1, 1978.

Fountain, D., Airborne electromagnetic systems—50 years of development. Exploration Geophysics, **29**, 1–11, 1998.

Frischknecht, F.C., 1967, Fields about an oscillating magnetic dipole over a two-layer earth, and application to ground and airborne electromagnetic surveys: Quarterly Colorado School of Mines, 62, 1–326

Grant, F.S. and West, G.S., Interpretation Theory in Applied Geophysics, McGraw-Hill, 1984.

Huang, H. and Fraser, D.C., The differential parameter method of multifrequency airborne resistivity mapping. Geophysics, **61**:1, 100–109, 1996.

Reid, J.E., Worby, A.P., Vrbancich, J. and A.I.S. Munro. Shipborne electromagnetic measurements of Antarctic sea-ice thickness. Geophysics, **68**:5, 1537–1546, 2003.

Palacky, G.J., The airborne electromagnetic method as a tool of geological mapping, Geophysical Prospecting, 29, 1981.

Palacky, G. J., Resistivity Characteristics of Geologic Targets. Electromagnetic Methods in Applied Geophysics, Society of Exploration Geophysics, 1988.

Sacramento Area Flood Control Agency, http://www.safca.org/Images/Maps/SacramentoRiverFloodControlSystem.pdf, 2018.

Siemon, B. "Electromagnetic methods—frequency domain", in Groundwater geophysics—a tool for hydrogeology, ed. R. Kirsch, Springer-Verlag, 2006.

Sorensen, J. C., and Chowdhury, K. 2010. Levee Subsurface Investigation Using Geophysics, Geomorphology, and Conventional Investigative Method. Proceedings of 30th Annual USSD Conference, Sacramento, California, April 12–16, 2010.

Spies B.R. 1989. Depth of investigation in electromagnetic sounding methods. Geophysics 54, 872–888.

Stanley H. Ward, (1967), 2. Part A. Electromagnetic Theory for Geophysical Applications, General Series : 13–196

Ward, S.H. and Hohmann, G.W. 1988, Electromagnetic theory for Geophysical applications, in Nabighian, M. N. Ed., Electromagnetic methods in applied geophysics, Vol. 1: Society of Exploration Geophysics, 131–312.

West, G.F and Macnae, J. Physics of the electromagnetic induction exploration method, in Electromagnetic methods in applied geophysics, Ed.M Nabighian, Society of Exploration Geophysicists, 1991.

Printed in the United States
By Bookmasters